ADVANCED TECHNOLOGY AND ITS MULTIFACED INVOLVEMENT IN EXPLOITATION OF OCEAN RESOURCES

THE EDITORS

A. B. Thakur, CE, FIE (India) Former Development Adviser, Shipping, Ministry of Surface Transport, Government of India. Founder Director, National Ship Design and Research Centre, Visakhapatnam.

Cmde Rajan Mathew joined the Indian Navy in 1970 after Graduation in Mechanical Engineering from Kerala University. He is a post-graduate in Marine Engineering from Institute of Armament Technology, Pune and in Energy Studies from I.I.T. Delhi. He is presently Director of Systems (Engineering) in Naval Headquarter.

B. S. Makhija is a Consulting Engineer, retired as Principal Surveyor of Ships, Hong Kong Government. He is the author of several papers and was awarded the Marine Engineering Division Gold Medal by The Institution of Engineers (India) for his paper entitled Humanising the Work Environment.

Dr. S. N. Suman, CE, FIE (India), A Chartered Engineer and Fellow of Institution of Engineers (India). Former Director (Mech.) in Ministry of Surface Transport, Government of India. He has many technical papers to his credit. He is also a Member of Institution of Plant Engineers and Institute of Standards. Presently he is a Consulting Engineer in Mech./Marine Engineering.

THE INSTITUTION OF ENGINEERS (INDIA)

The Institution of Engineers (India), IEI is the first professsional body of engineers founded in India in 1920 and incorporated by Royal Charter in 1935, to promote and advance the art and science of myriad fields of engineering and technology.

IEI represents India in several international bodies such as, the World Energy Council, World Mining Congress, World Federation of Engineering Organisations, Commonwealth Engineers Council, International Council of Aeronautical Studies and several others.

As the Indian Member Committee of the World Energy Council since inception, The Institution of Engineers (India) was privileged to host this International Conference on Energy-Asia Energy Vision 2020—particularly when several countries including India, have made political decisions, towards a free market economy and globalisation of Industry and commerce.

ADVANCED TECHNOLOGY AND ITS MULTIFACED INVOLVEMENT IN EXPLOITATION OF OCEAN RESOURCES

Proceedings of the 12th National Convention of Marine Engineers, organised by The Institution of Engineers (India), Delhi State Centre, during February 12th & 13th, 1999 at New Delhi

Editors
A.B. THAKUR
Cmde R. MATHEW
B. S. MAKHIJA
Dr. S. N. SUMAN

NEW DELHI
February 12th & 13th 1999

Published for

DELHI STATE CENTRE
THE INSTITUTION OF ENGINEERS (INDIA)

by

CONCEPT PUBLISHING COMPANY, NEW DELHI

ISBN 81-7022-783-6

First Published 1999

Published and Printed by
Ashok Kumar Mittal
Concept Publishing Company
A/15-16, Commercial Block, Mohan Garden
New Delhi 110059 (India)

Phones: 5649024, 5648039
Fax: 091–(11)–5648053

PREFACE

Considering the 6000 km long coast line, the size of the Maritime Economic Zone of India is considerably large. The prospect and potential for successful commercial exploitation of the vast resources lying there in, therefore, is a unique challenge, both in terms of investment as well as technological involvement. The involvement of technology in this regard is really vast and call for intensely integrated multi-disciplinary engineering and technological inputs. The all embracing nature of the involvement could be gauged from the facts, that a number of important economic ministries of the Government of India, such as Ministry of Petroleum & Natural Gas, Energy, Surface Transport, Department of Ocean Development, Department of Fisheries etc. are already directly or indirectly involved in this sector. Considering the liberalised economic environment and some of the surveys/studies carried out in this regard, it may not be difficult to attract the necessary investment and technological inputs for more intensive exploitation of these resources.

In view of the economic importance to the country as well as national and international involvement in terms of investment and technology, the Delhi State Centre, Marine Engineering Division of the Institution of Engineers (India), consider it appropriate to hold a National Seminar along with its 12th National Convention of Marine Engineers on Advanced Technology and its multifaced involvement in exploitation of ocean resources. The deliberations/participation of eminent personalities, authors, engineers and other professional involved in the related field. The proceeding contains fifteen papers I trust that it will help to project the problem and prospects of various aspects of the involvement in its proper perspective.

A. B. THAKUR
Chairman,
Organsing Committee

CONTENTS

IV. MAN POWER AND TRAINING

I

GENERAL

MARITIME TRAINING—STCW 1995 IMPLICATIONS

1

Debabrata Bandyopadhyay
CEng (Lond) FIMarE FIE
Faculty, Marine Engineering and Research Institute, Calcutta

ABSTRACT

Maritime safety is one of the prime concerns in maritime states all over the world. Various ship casualties have been studied and researches continued to ascertain causes thereof, which finally brought about either amendments to existing safety regulations or adopt some other modified instruments which the ship has to comply with.

In spite of stricter regulations formulated by various national and international bodies, including the IMO (International Maritime Organisation), ship losses and casualties did not reduce to the extent expected.

Stricter regulations endeavoured modifying HARDWARE parts of the ship. Attention was therefore diverted to improving the SOFTWARE parts, namely the management and those who man the ships.

However by 1978 STCW (Standards of Training Certification and Watchkeeping) was no longer seen adequate to cope with the current technological advances. The IMO and its Maritime Safety Committee accordingly revised the STCW convention in the conference held at London in June 1995.

The revised STCW is quite logical and makes it very convenient for the provisions to be easily identified with reference to regulations and codes. Tacit acceptance procedure has been adopted for worldwide implementation on a speed track. STCW 95 provides standards as per functional approach adopted in shipboard operations. The standards are stipulated for each of the three functional levels.

- MANAGEMENT LEVEL — comprising four senior officers on board Captain, Chief engineer, Chief officer and Second engineer.
- OPERATIONAL LEVEL — comprising navigating, radio and engineer watchkeeping officers.
- SUPPORT LEVEL — comprising ratings of all above.

Each level requires to be trained adequately as per tabled format given in Chapters II, III, IV and VI in STCW 95 code.

The objectives of STCW revised convention are:

(a) that the training institutions around the world would effectively comply with enhanced standards.

(b) that the seafarers themselves are adequately trained and kept abreast of technological developments.

(c) that the certificates are issued only to those who have demonstrated in a clear and unambigous manner that they are competent and deserve to be holders of such certificates at all time.

On the basis of guidelines given in the STCW codes and various IMO model courses, maritime institutions in India and abroad started restructuring training and assessment procedures at all three levels.

The subject paper deals with implications on modified education and training schemes, assessment certification and revalidation as per Merchant Shipping (STCW) Rules 1997, as well as on basic safety management (ISM code) implementation.

"Running a ship overseas with maximum safety measures" is a topic which has been thought of since long. Various ship casualties have been studied and researches continued to ascertain causes thereof, which finally brought about either amendments to existing safety regulations or adopt some other modified Instruments which the ship has to comply with. To cope up with modern technological advancement the ship has to be constructed more safe and installed with latest equipments, the net idea being one and only one i.e. to create a safe ship.

It is undoubtedly true that stricter regulations and modified equipments are necessary at par with the development of science. Over and above, for control of same, strict vigilance is also required to be observed by prospective shipowners and the Administration.

One must not forget the key factor resulting in ship disasters most of the time, which is HUMAN ERROR and POOR CONSCIOUSNESS IN SAFE OPERATION OF SHIP. Ships are operated and maintained by people sailing with competency/proficiency certificates awarded to them by the Administration as per internationally accepted policies. Therefore, to achieve safety consciousness and competency standard adequate training is necessary at all stages so that seafarers develop safety culture and maintain it with the changing trend of modern times.

IMO (International Maritime Organisation) realised the necessity of comprehensive maritime training and therefore brought about STCW (Standards of Training Certification and Watchkeeping) convention 1978 which came into force with effect from 1984 in a few maritime states. Others could not comply due to various administrative and infrastructural problems. It was agreed that certified watchkeeping personnel would undergo approved safety courses like First Aid at Sea, Survival at Sea, Basic and Advanced Firefighting, Tanker Safety (Dangerous Cargo Endorsement) etc. for the certificate to remain validated for seagoing service. Is this enough for safety of ships? If yes, then why did IMO have to face global criticism in the aftermath of spate of accidents that occurred in various parts of the world in 1992-93, long after STCW 78 came into force. The criticism resulted in branding the IMO as slow and by implication, inefficient. IMO and its Maritime Safety Committee (MSC) and the council, therefore demonstrated in the STCW conference held at London on June 1995 that they are not inefficient, by bringing about changes in the convention, called STCW 95 and advised all maritime states all over the globe to take speedy action for its implementation.

The task of implementation of the convention requirements in full is not easy. Realising this at the onset of conference, IMO's STCW sub-committee approved that a group of consultants supported by experts designated by Maritime Training Institutes from around the world should be given the task of preparing the amendments to STCW 1978 convention.

STCW REVISED CONVENTION 1995

The MSC agreed that the revised convention should be restricted to broad principles and legal arrangements. And that technical details should be placed in STCW code for mandatory application, as appropriate, by incorporation of an appropriate reference in the convention itself.

The draft of the revised convention consisted of :-

1. Regulations
2. STCW code
3. Conference Resolutions

Traditionally the IMO Instruments enter into force after a fairly long period when the requisite number of maritime countries, representing a defined percentage of world's tonnage, have ratified the convention. However, in the case of STCW revised convention Tacit Acceptance procedure was adopted and the acceptance process was kept on the so called 'Speed Track'.

As stated by Mr. William. A. O'Neil, the Secretary General of IMO, the effective implementation of the standards which the revised convention contains will be achieved by ensuring

- that the training institutions around the world effectively comply with enhanced standards,
- that the seafarers themselves are adequately trained and kept abreast of technological developments, and that the certificates are issued only to those who have demonstrated in a clear and unambiguous manner that they are competent and deserve to be holders of such certificates.

True enough, the revised STCW convention is quite logical and makes it very convenient for the provisions to be easily identified with reference to Regulations and Codes. Eight chapters of the revised convention consist of following :-

1. General provisions
2. Master and deck department
3. Engine department
4. Radio communication and Radio personnel
5. Special training requirements for personnel on certain types of ships
6. Emergency, occupational safety, medical care, survival functions
7. Alternative certification
8. Watchkeeping

Chapters II to VI contain in a Chart form the training necessary to be developed for various certifications of nautical, engineering, radioelectronics officers and ratings.

STCW 78 stipulated the 'minimum knowledge required for certification' and by the time worldwide adoption was secured, the standards would have been very old and not at par with modern days. The revised convention goes much beyond this and provides for FUNCTIONAL APPROACH in shipboard operations, under following heads :-

1. Competence
2. Knowledge, understanding and proficiency
3. Method of demonstrating competence
4. Criteria for evaluating competence

These standards are stipulated for each of the three FUNCTIONAL LEVELS such as

1. **Management level** comprising of four senior officers, Captain, Chief engineer, Chief officer and Second engineer .
2. **Operational level** comprising of navigating and engineering watchkeeping officers.
3. **Support level** comprising of ratings for above.

Requirements for each level are elaborated in a tabular form in Chapters II, III, IV and VI of the STCW code.

The Conference STCW 95 adopted

- the amendments to the annex to the International Convention of STCW 78 together with **resolution 1 on adoption of the amendments which constitutes Attachment 1 to the final act.**

  The seafarers training, certification and watchkeeping code together with **resolution 2 on adoption of the code which constitutes Attachment 2 to the final act.** The conference also adopted the following resolutions in Attachment 3 to the final act :-

Resolution 3 — TRANSITIONAL PROVISIONS

The conference agreed to bring into force the proposed amendments on 1st February 1997, with a grace period of five years, for those who had already commenced their training under existing provisions, which means, until 31st January 2002, continued to revalidation or renewal of certificates issued prior to 1st February 1997.

Resolution 4 — TRAINING OF RADIO OPERATORS FOR GMDSS

In order to comply with the requirements of Radio communication and Radio personnel (Chapter IV) the proposed GMDSS operator's course needs to be started at the earliest. GMDSS worldwide is due to be in force with effect from 1st February 1999 which is much before the terminal date of other transitional certificates. Existing radio operators therefore need to be upgraded to avoid a sudden vacuum in the industry.

Resolution 5 — TRAINING IN CRISIS MANAGEMENT AND HUMAN BEHAVIOUR FOR PERSONNEL SERVING ON RO-RO PASSENGER SHIPS

Crisis management and human behaviour training is considered to be essential for key personnel at management level on ro-ro passenger ships and should include the following elements in training :—

1. Human behaviour and responses,
2. Optimizing the use of resources,
3. Development of emergency plans,
4. Leadership skills,
5. Stress handling and communication.

Resolution 6 — TRAINING OF PERSONNEL ON PASSENGER SHIPS requires specialisation in certain areas apart from essential training applicable for other ships.

Resolution 7 — MONITORING THE IMPLICATIONS OF ALTERNATIVE CERTIFICATION

Chapter VII allows the Administration, if they so desire to have alternate arrangements not solely dependent on conventional method.

While introducing this alternate certification scheme, it should not in itself be used for

- reduction of crew
- lowering the dignity of the crew or deskill them
- justifying the assignment of the combined duties of the engine and deck watchkeeping officers to a single alternate certificate holder
- the person in command shall be designated as master and the legal position and authority of the master and others shall not be adversely affected
- lowering safety standards, the minimum requirements as per Chapters II and III. Regulations to specify clearly the requirements of education and training for alternate certification.
  For watchkeeping certification, candidates shall have approved sea-going service of not less than 12 months, out of which six months shall be when performing engine room duties. And when navigating function is to be included, service for six months shall be performed on bridge watchkeeping duties under supervision.
- where the service in capacities in positions other than management level is to be considered, the candidates shall be required to show service at operational level of 12 months on deck and 12 months in engine room
- when service at the management level is to be performed, the candidates shall be required to serve for 24 months, as specified in A-II/1 and/or A-III/1 & A-III/2 .

Resolution 8 — PROMOTION OF TECHNICAL KNOWLEDGE, SKILLS AND PROFESSIONALISM

The Administration should ensure that shipowners

- establish criteria and processes for the selection of personnel exhibiting the highest practicable standards of technical knowledge, skills and professionalism
- monitor the standards exhibited by ship's personnel in performing their duties
- encourage all officers to participate actively in the training of juniors
- monitor carefully and frequently review the progress made by juniors in their acquisition of knowledge and skill during their service on board ships
- provide refresher and updating training at suitable intervals as may be required
- take all appropriate measures to encourage pride of service and professionalism.

Resolution 9 — DEVELOPMENT OF INTERNATIONAL STANDARDS OF MEDICAL FITNESS

IMO is in process of developing such standards for seafarers in co-operation with the International Labour Organisation and the World Health Organisation.

Resolution 10 — TRAINING OF MARITIME PILOTS, VESSEL TRAFFIC SERVICE PERSONNEL AND MARITIME PERSONNEL EMPLOYED ON MOBILE OFFSHORE UNITS

Resolution 11 — PROMOTION OF TECHNICAL CO-OPERATION

This should be conducted at an intergovernmental level to assist those states not yet having adequate expertise or facilities for providing such training and experience to implement the revised STCW convention requirements.

Resolution 12 — CONTRIBUTION OF WMU IN THE ENHANCED STANDARDS OF TRAINING

The Conference urges the IMO to utilise the resources and expertise of WMU to transfer maritime education and knowledge wherever needed, particularly in third world developing countries. It should encourage and, support WMU to assume a leading role for the promotion and implementation of a network of advanced maritime training establishments.

Resolution 13 — REVISION OF MODEL COURSES PUBLISHED BY IMO

The Organisation to take steps and revise and update those model courses which provide guidance on implementation of training and assessment provisions of the STCW convention as amended.

Resolution 14 — PROMOTION OF PARTICIPATION OF WOMEN IN MARITIME INDUSTRY

Governments of various flag states are required to give special consideration to securing equal access by men and women in all sectors of the maritime industry. The role of women in the seafaring profession to be highlighted to promote their greater participation in the maritime training at all levels.

TRAINING SCHEME AS PER STCW AS AMENDED IN 1995

In India major maritime training institutions are :—

1. Marine Engineering and Research Institute Calcutta
2. Marine Engineering and Research Institute Mumbai
3. L.B.S. College of Advanced Maritime Studies and Research Mumbai
4. Training Ship Chanakya Mumbai

1. At M.E.R.I Calcutta, presea training is being given, complying with Table A III/1. Additional academic and practical studies finally lead to the award of a First Degree in marine engineering.

   INTAKE LEVEL :- (i) Pass in 10+2 course with English, Physics, Chemistry, Mathematics, (ii) Pass in J.E.E conducted by Indian Institute of Technology, and (iii) counselling by competent authority.

   COURSE DURATION :- Four academic years divided in eight semesters, adequately covering a Degree syllabus approved by All India Council of Technical Education.

   Approximately 120 candidates pass out from this Institute and every year and acquiresix months ship-board experience to become eligible for Class IV examination. Entire training is being conducted under documentation of approved T.A.R . book (Training and Assessment Record) prepared as per requirements of STCW 95 Code Part A. Level of competence is judged at regular interval throughout the tenure of training and grading of individuals candidate assigned thereto in his T.A.R Book

2. In M.E.R.I. Mumbai, presea training course, complying with mandatory requirements of the Code, Table A- III/1, for graduate engineers, is being conducted.

   INTAKE LEVEL :- Pass in Degree in Mechanical Engineering from any recognised university in India.

   COURSE DURATION :- 6 months marine engineering training in class/lab/ workshop/simulator and 6 months training aboard merchant ships.

   These candidates are normally selected by a shipping company as trainee marine engineer who undergo academic training in M.E.R.I. Mumbai and on board training in ships of the concerned shipping company under guidance of competent trainers, all being documented in T.A.R. books under STCW.

3. In L.B.S. college following advanced courses are being conducted :—

   (a) Post-sea courses as required under STCW as amended in 1995,
   (b) Preparatory courses for competency examinations,
   (c) Extra Master's (India) preparatory course,
   (d) Extra First Class Marine Engineer's (India) preparatory course.

4. In T.S. Chanakya pre-sea training in given to nautical candidates as per STCW 95 Code, Table A-II/1.

   INTAKE LEVEL is the same as in M.E.R.I., Calcutta. The course extensively covers requirements of STCW 95 under T.A.R book documentation, and finally a BSc degree in nautical science is awarded to the passing out candidates by the Mumbai University.

OTHER MARITIME TRAINING INSTITUTIONS IN INDIA	LEVEL OF INSTRUCTIONS
1. Academy of Maritime Education and Training Chennai.	An ISO 9002 pre sea training centre.
2. BPT-FOSMA Seafarers Training Centre, Mumbai.	Rating training.
3. Cadet Academy Great Eastern Shipping Mumbai.	Pre-sea nautical cadets training.
4. Cochin Shipyard Marine Engineering Training Institute.	Pre-sea training Engg graduates.
5. Indian Institute of Port Management Calcutta.	Post-sea Courses.
6. FOSMA Maritime Institute and Research Organisation.	Nautical & Marine Engg pre-sea and post-sea courses.
7. Indian Maritime College Chennai.	Radio operators course.
8. National Institute of Port Management Chennai.	Pre-sea and post-sea courses, Nautical / Marine Engg.
9. Orissa Maritime Academy, Paradeep.	Rating training.
10. Sea Explorers Institute, Calcutta.	Life saving and first aid courses.
11. Seacom Marine College, Calcutta.	Radioelectronics, GMDSS course.
12. Maritime Training Institute, SCI, Mumbai	Pre-sea and post-sea courses. Nautical/Engg.
13. Calcutta Technical School, Calcutta.	Pre-sea Marine Engg Diploma
14. Tolani Maritime Institute, Pune	Pre-sea Marine Engg Degree
15. Chidambaram Marine Institute, Chennai	Pre-sea training Engg graduates
16. International Maritime Institute, New Delhi	An ISO 9002 per-sea training centre. National Marine Engg.

Under the guidance of STCW amended requirements, the Directorate General of Shipping has been encouraging inspiring various shipping companies and workshops to build up their own training centres to meet the increasing demand of maritime personnel in the industry.

The requirements of STCW 95 have been stipulated in a consolidated form in the Indian Merchant Shipping (STCW) Rules 1997. Marine Institutions imparting training to navigating and engineering cadets are required to prepare Training and Assessment Record books (T.A.R) for Institutional as well as On-board training separately, as per Chapters II and III of the convention. Every candidate must possess T.A.R books which give his training and assessment record, right from entry. Before appearing in any examination for certificate of competency applicable to his rank at operational level, he must submit T.A.R books to examining authorities, duly endorsed by appropriate authorities of training centres and on ships certifying his competence towards preliminary assessment. The final assessment will be carried out by designated assessors at the assessment centre.

ENGINEER'S QUALIFICATIONS AND EXPERIENCE

Each candidate for marine engineer officer (MEO) Class IV, Class II, Class 1 or Class III Certificate of Competency (COC) shall have

1. Initial training as per chart 2A, 2B (see Annex),
2. Qualifying sea service as applicable in table coc and successfully completed an approved preparatory course (Table 2A, 2B, 2C) and a simulator course for Class IV, Class II and Class I as appropriate,
3. Passed the examination for the respective class, specified in table COC (capacity column)
4. Undertaken an approved engineering management course for MEO Class I.

A candidate for Extra First Class MEO Certificate (India) shall—

1. Hold a MEO Class I Motor or Steam or Steam & Motor Combined Certificate,
2. Have satisfactorily completed an approved preparatory course for Extra First class (12 months),
3. Passed the examination—Parts A, B and C. (Part A & B constitute academic subjects and Part C dissertation).

REVALIDATION OF COCs

Every master, engineer, deck officer and radio operator holding a COC issued by the Administration, who is serving at sea or intends to return to sea after a period ashore shall, in order to continue to qualify for sea-going service, be required at intervals not exceeding five years to:

1. meet standards of medical fitness as per ILO convention, and
2. establish continued professional competence as under :-
   either, approved sea-going service performing functions appropriate to the COC held for a period of at least one year in total during the preceding five years
   or, passing an approved test
   or, successfully completing an approved course or courses.

The refresher and updating courses as required by Regulation I/11 of STCW 95 should be approved by the Administration concerned and include ondate technological advancement in ship's operation and maintenance along with relevant amendments in national and international regulations concerning safety of life at sea, loadline and protection of marine environment.

SIMULATOR BASED TRAINING

Section A-I/12 of STCW 95 Code provides guidelines for standards governing the use of simulators in maritime training. Trainees must have adequate familiarisation time on the simulator and its equipment. Exercises are to be effectively monitored, supported as appropriate, by audio and visual observation of training activity together with pre and post exercise evaluation reports. The training must be capable of simulating the operational capabilities of the shipboard equipment concerned, to a level of physical realism, appropriate to training objectives.

ISO/ISM IMPLICATIONS

Inspite of stricter regulations formulated by various national and international bodies including the IMO, ship losses and casualties did not reduce to the extent expected. This is perhaps because improvements in the HARDWARE part of ships were not accompanied by those concerning management. Attention was therefore diverted to improving the SOFTWARE part, namely the management personnel and those who man the ships. These attempts have resuted in Safety Management Code, known as ISM Code, as well as adoption of Quality Management in accordance with the ISO 9000 series.

The ISM code became mandatory for all types of ships in a phased manner with effect from 1st July 1998, while the ISO 9000 on quality management is entirely voluntary. Quality management is, however, necessary for long term survival in the highly competitive market. It is worth remembering that unlike a defective product, a deficient service once provided can not be retrieved.

The basic premise of both the ISM and ISO systems is to provide measures that decrease the risk of making mistakes. These are applicable, therefore, to both Industry as well as Training houses.

Four major safety conventions are enforced in shipping namely, **Loadline, Solas, Marpol and STCW**. Statutory certificates issued under these conventions need to remain validated at all time for smooth trading of the vessel concerned. To monitor all these convention requirements alongwith a proper safety management system , each ship will be issued Safety Management Certificate (SMC) valid for five years, subject to Annual and Intermediate audits/surveys.

Under the STCW convention, independent evaluation of knowledge, understanding, skills and competence acquisition in Training, requires an Audit survey to be conducted to verify

- that all internal management control, monitoring measures and follow up actions comply with planned arrangements and documented procedures and are effective in ensuring achievement of the defined objectives,
- that the results of each independent evaluation are documented and brought to the attention of those responsible for the area evaluated,
- that timely action is taken to correct deficiencies if any.

The ISM code is mandatory for virtually every ship in the world. Shipping companies which do not have a safety management system in place would be in violation of the Safety of Life at Sea (SOLAS) convention and could find themselves out of business. The ISM Code also requires ships to be manned by crew with approved standards of training and certification (STCW revised) .

The greatest return expected through implementation of the ISM code and the safety management system will be the reduction of accidents and eventual increase in efficiency, through promotion of the safety culture.

NEAR COASTAL VOYAGES (NCV)

Section I/3 of the revised STCW convention provides clearer principles for declaring near coastal waters.

Mandatory maritime training and experience required for NCV certification are relaxed considerably. It further requires that seafarers on ships flying the flag of another country shall also be permitted similar relaxation.

It had been a practice for quite sometime to grant certificate of competency to navigating officers including masters, on HOME TRADE and/or FOREIGN GOING ships, depending on their basic qualifications, training and experience, but there was no clear distinction for engineers certificates till now.

To provide for viable manning on vessels plying near coastal voyages the Government of India, like all other flag states, propose to finalise a comprehensive scheme of training under guidelines of STCW 95 and send it to the Secretary General IMO for approval. This would certainly benefit the huge coastal fleet of India to achieve good qualified manning strength.

ER ratings on watch, sea going engine drivers, diploma holders in mechanical or/and electrical engg, second class engine drivers, marine fitters etc. are required to undergo a certain amount of sea service (depending on qualifications) after completion of four basic safety courses, namely—

1. Personal survival technique (PST)
2. Basic fire fighting
3. Elementary first aid
4. Personal safety and social responsibility followed by three months course covering AIII/1 read with A I/3.

These are to be followed by three advanced safety courses, namely—

1. Advanced fire fighting
2. Proficiency in survival craft and rescue boats, other than fast rescue boats
3. Medical first aid Upon completion of three courses, followed by six months sea service (onboard Training) authenticatd by T.A.R book documentation, they will be entitled to appear for MEO Class IV (NCV), written and oral examination.

The Class IV (NCV) certificate holders would require further twelve months sea service, followed by four months Institutional course, covering A III/3, as applicable for sections III/3 and I/3 of STCW 95. Thereupon, such candidates can appear for MEO Class III Part A and Part B examinations, passing of which would entitle them to serve in the rank of 2/E/O (NCV).

For upgradation to C/E/O (NCV) candidates have to perform further twelve months sea service and subsequently undergo training for three months covering Code A III/3, read with A I/3 of STCW 95. They can then sit for the written and oral examination by the Administration for certification as C/E/O (NCV)

OR after passing aforesaid examination sail on offshore vessels of less than 6000kw as 2/E/O. One year of such sea service will enable the candidate to appear for an oral examination by the Administration and, if successful, the candidate will be certified as C/E/O for less than 6000KW OSVs.

OR a candidate holding C/E/O (NCV) certificate may, if applicable in his case as per Rules, undertake for six months a course covering Code A III/1 & Code A III/2 (selected topics practically essential for safety) of STCW 95 and pass MEO Class IV part A examination for ocean going ships.

STCW 1995 stresses in levels a quality training system. The field of application of the quality standards covers the Administration of certification system, all training courses and programmes, examination and assessment carried out by or under the authority of training organisation and the qualifications and experience required for Instructors and Assessors, having regard to the policies, systems, controls and internal quality assurance reviews, established to ensure achievement of the defined objectives.

TABLE 1.1 : **COC Certification in Engineering Department**

Sr. No.	*Capacity*	*(KW) Propulsive Power*	*STCW regn.*	*Minimum months watchkeeping time*	*Total months sea service*
1.	OICEW (MEO IV)	750 or more	III/1	4 Steam or Motor	6
				8 Steam + Motor	8
2.	2EO (MEO class II)	3000 or more	III/2	9 Steam or Motor	12
				18 Steam+Motor	18
3.	CEO (MEO class I)	3000 or more	III/2	12 Steam or Motor	36
				18 Steam+Motor	36
4.	OICEW MEO class IV(NCV)	750 to 2999	III/1 read with I/3	6 with T.A.R book	
5.	2EO MEO class III (NCV)	750 to 2999	III/3 read with I/3	12 (sea service)	
6.	CEO MEO class III (NCV)	750 to 2999	III/3 read with I/3	12 (sea service)	
7.	Ratings on engine watch	750 or more	III/4	24	6 (T.A.R)

NCV — Near Coatal Voyage

OICEW — Officer Incharge Engineering Watch

WMU — World Maritime University

ANNEX 1

TABLE 1.2A : Entry qualification and initial training requirements for joining a ship as an assistant engineer officer (Reference STCW-95 Code A-III/1)

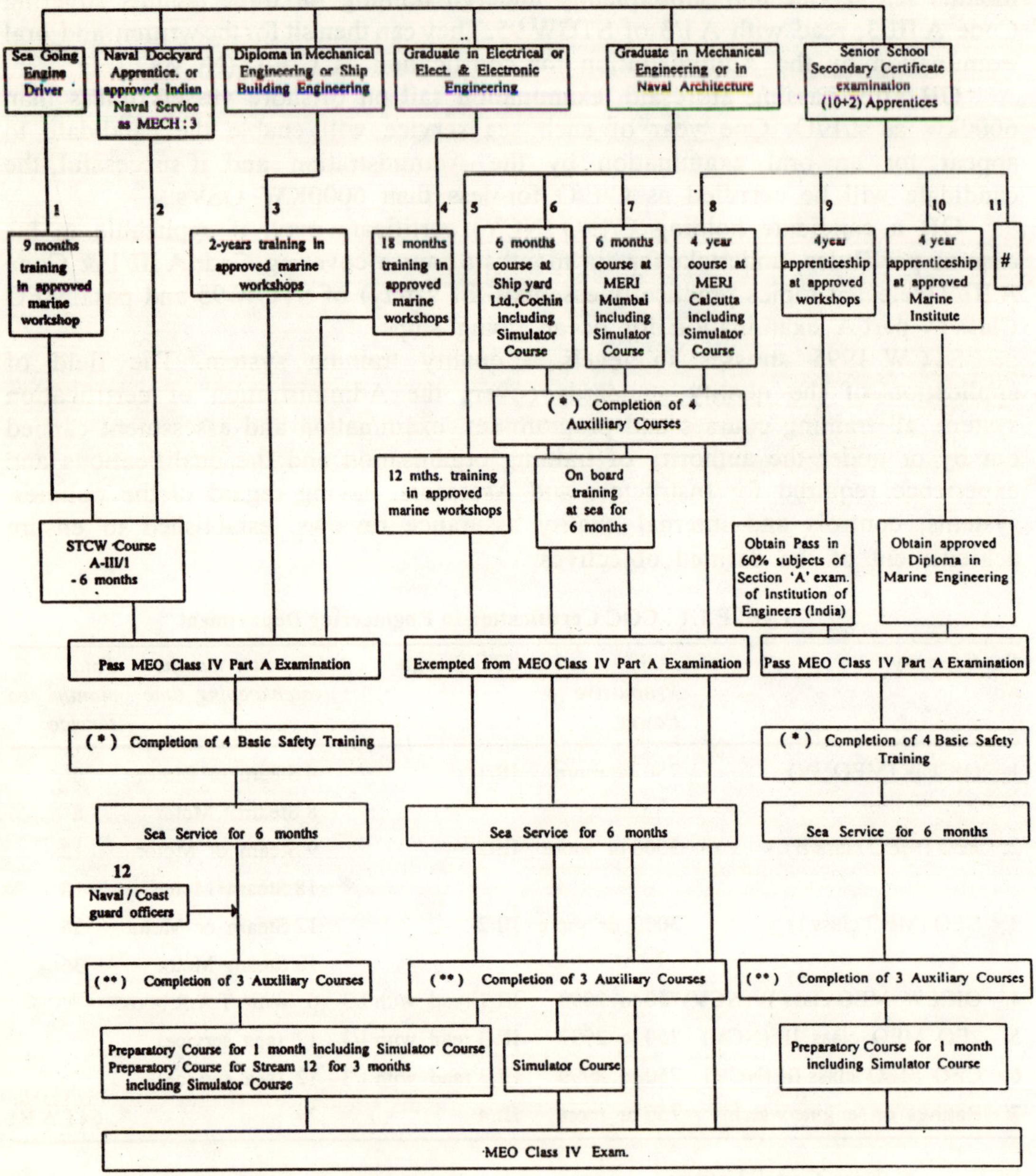

(*) - (i) Personal Survival Techniques as per regulation.
(ii) Basic Fire Fighting.
(iii) Elementary First - Aid.
(iv) Personal Safety & Social Responsibility.

(**) - (i) Advanced fire fighting.
(ii) Proficiency in Survival Craft & rescue boats other than fast rescue boats.
(iii) Medical First aid.

ANNEX 2

TABLE 1.2B : **Alternative training scheme (30 months)**
(Reference STCW - 95 Code A-III/1)

11

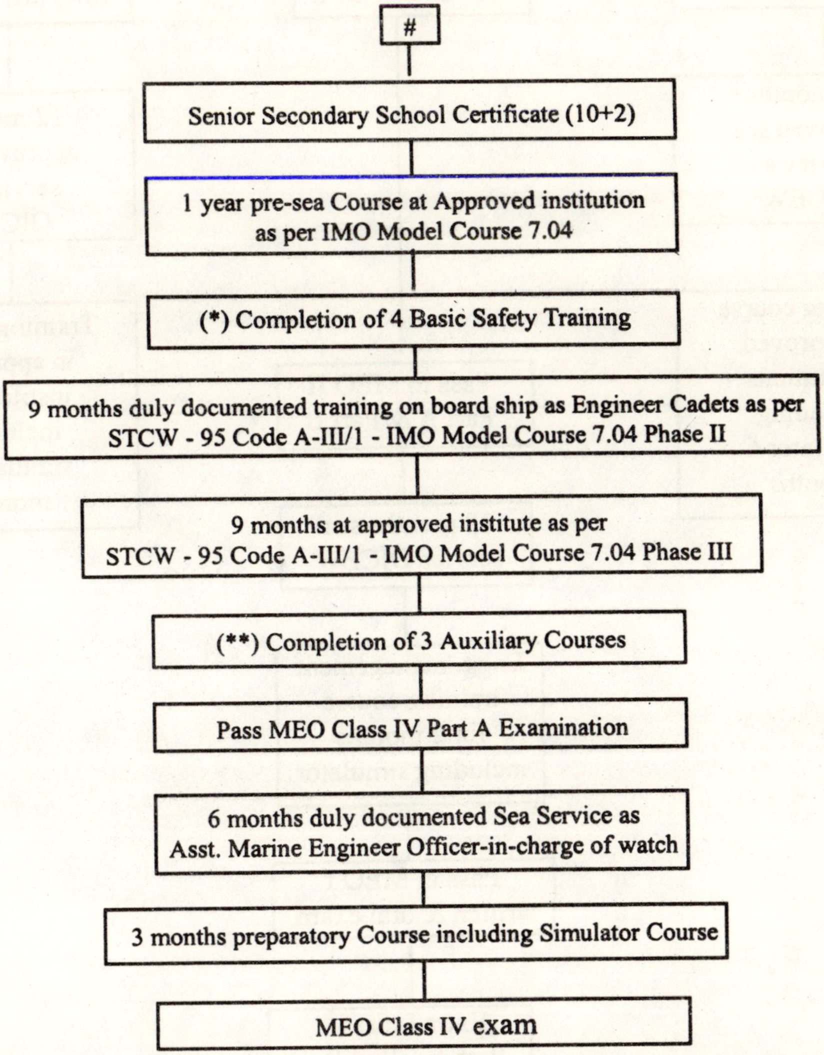

(*) - (i) Personal Survival Techniques as per Regulation.
(ii) Basic Fire Fighting.
(iii) Elementary First - Aid.
(iv) Personal Safety and Social Responsibility.

(**) - (i) Advanced Fire Fighting.
(ii) Proficiency in Survival Craft and Rescue Boats other than Fast Rescue Boats.
(iii) Medical First Aid.

ANNEX 3

TABLE 1.2C : **Progression from MEO Class IV to Extra First Class level**

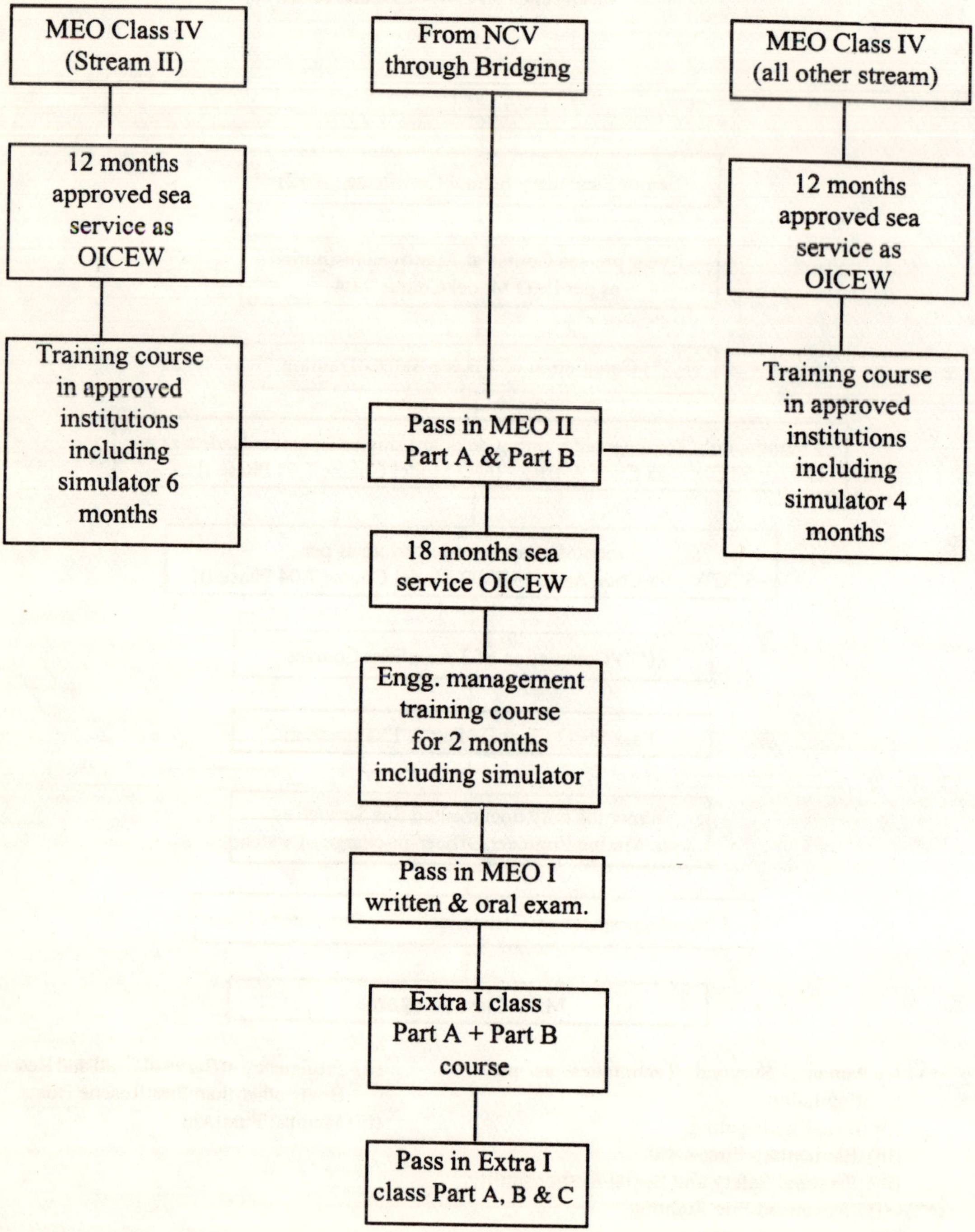

SIMULATORS FOR TRAINING

2

Lt. Manish Vig and Lt. S. Kapahi
INS, Ranvir

INTRODUCTION

Ever since the days of oars and sails, there has been a steady progress in the level of technology used in marine systems. The quest for higher power densities, increased efficiency and rapid responses, supported by continuous technological developments, has resulted in design and development of machinery configurations of ever increasing sophistication and also ever decreasing tolerance limits. This makes them highly susceptible to damage due to maloperation. Serious damage to the machinery is certain in the event of operator error, unless corrective action is taken instantly. Hence, safe and correct operation of these vessels calls for well-trained and competent operators. Use of live equipment for training is obviously ruled out in view of the heavy penalties that may be imposed in terms of cost and down time.

With advances in automation, the number of operators on board have gradually shrunk in arithmetic proportions, but at the same time the required level of knowledge and skill on the part of the operator has grown in geometric proportions. This has necessitated a harmonic increase in the level of training inputs, both in operation and in maintenance aspects.

The time honoured methods of training are fast becoming obsolescent. More sophisticated training aids compatible with the training and performance objectives need to be employed. It is here that simulators have a distinct place of their own and can become effective tools in meeting the stringent demands imposed on the level of operator training. Other new methods of training, like multimedia based training, may also supplement the training, serving to improve the operator's theoretical knowledge.

This paper discusses the role of simulators, for training of operators and also gives a brief insight into the use of simulators in the Indian Navy.

WHAT IS SIMULATION

It may be useful to discuss at the outset what is meant by Simulation. In simple terms, simulation can be thought of as a representation of reality in mathematical terms and realisation of its behaviour in, as far as possible, a realistic environment. For example, one can describe the trajectory of a missile by a mathematical equation, then subject the equation to certain tests represented by other mathematical terms to see the effect on the missile under those conditions.

A full scope simulation of the propulsion plant of a ship for operator training is a complex task. The mathematical modelling of a propulsion plant is concerned with a large number of physical variables, such as temperature, pressure, mass flow, flow rates, enthalpies, vibrations, energy storage etc. It is advantageous and cost-effective to provide as many simplifications as possible within the limits of steady state and transient accuracy. The models governing such a large number of variables are complex mathematical equations, which require a detailed analysis, and step by step data validation. An easier alternative to developing mathematical models is the use of look-up tables. Look-up tables are simple data-files, which contain a pre-recorded data from a live plant. The simulation software is therefore significantly simplified. The task can be further simplified to a certain extent by adopting a modular approach. The various modules developed may be easily transferred to simulate other systems as well.

MERITS OF SIMULATORS

Training in the Navy is a costly, labour intensive and time-consuming process. Training an inexperienced operator on actual equipment can lead to damage of equipment thereby resulting in high repair cost and/or down time. A method for offsetting this hazard is available through use of simulators for training. There are a number of advantages that demand introduction of simulators for training in order to obtain first rate operators of the sophisticated equipment and machinery on-board.

(a) No Risk to Man /Machine

Training an operator on a simulator involves no risk at all. An inexperienced operator is likely to make mistakes, which can prove to be dangerous and costly on a live equipment. Whereas, mistakes made by the operator on a simulator would only add to the learning process.

(b) Nil Operating Cost

The cost of operating a simulator is negligible compared to the cost of operating live equipment specifically for training. The only cost involved in a simulator is the initial cost of the simulator, which is also quite low as compared to the cost of single live equipment. Therefore, if the training given on a simulator can prevent damage to single live equipment due operator inexperience, the simulator would have paid for itself.

(c) High Training Effectiveness

Simulators have proved to be the finest of all training aids currently available. Even multimedia applications fail to reach so close to reality as is achieved by the use of simulators. The use of simulators therefore leads to enhanced training effectiveness. The trainee is able to comprehend the theoretical lessons in a better way. This leads to a greater confidence of the operator.

(d) Training Without an Instructor

The unseen role of a simulator is that of an instructor. Using a few special features, a simulator can be effectively used for performance evaluation of trainees. Also, based on the trainee's evaluation, the simulator may be programmed, to suitably alter the simulation process, and certify the trainee for operation of live equipment only after he has cleared a certain level of performance on the simulator.

(e) Fault Simulation

This is a feature that is unique to simulators. Various emergencies that may occur on the actual equipment are inducted into the simulation process on a random or instructor-induced basis. This enables the operator to sharpen his reactions to various emergencies.

(f) Continuous Availability

Simulators can be operated continuously, year after year, with insignificant down time or performance deterioration. Real plants must necessarily be withdrawn from service for refitting and maintenance, overhaul at regular intervals, and consequent deterioration in performance level based on operating hours is inescapable.

(g) Quick Initialisation

A simulator can be initialised to any state of operation of a real plant within seconds, which is time consuming, undesirable and even next to impossible with real plants. For example, bringing a live steam plant to a desired operating point is time consuming. Repeated start up and shut down would be detrimental to the health of the plant and therefore undesirable, and such operations being impossible with nuclear plants.

(h) Time Scale Adjustment and other Features

The time scale of action on the simulator can be adjusted to a faster or slower pace than reality. This helps to cater for varying grasping capabilities of operators. In addition, demonstration of a process on a slowed time scale enables better understanding of the process. It permits recognition of the rapidly occurring changes without diminishing the logical validity. For example: the starting cycle of a gas turbine may be slowed down to any desired pace to enable a clear understanding of each change occurring. Further, there is total repeatability of information on a simulator, which cannot be accurately repeated on demand on a real plant. This together with record, replay, freeze and zooming facilities at any state point, allow for meaningful interaction between the trainee and the trainer.

(j) Pretrained Operators

Operator training can be started before the real plant is available, which apart from savings in time, certainly makes the initial operation of a new type of plant safer and easier, by making available a pool of trained operators as a part of the ship's commissioning crew.

(k) Performance Evaluation

The simulator may be programmed to evaluate each trainee on his performance on the simulator. Also, the trainee may be subjected to a Computer Based Training package.

CLASSIFICATION OF SIMULATORS

Simulators can be categorised in two distinct types, namely, 'Design simulators' and 'Training Simulators'. Design simulators usually cover only a small part of the process but in much more mathematical detail involving computations of theoretical or hypothetical situations, which may have no relevance from the operators point of view. Design simulators may run slower or faster than the real process, while training simulators must operate and respond in a manner identical to the actual plant and in that sense they must operate in real time. A design simulator may involve only a narrow range of conditions for detailed study, while a training simulator must simulate from 'cold' condition the start up actions up to 'shut down' conditions. A design simulator usually revolves around the major process, while a training simulator is required to cover every process in both the main and the auxiliary systems with which the plant is provided, including associated display, alarm, telegraph order sensing, measuring and transmitting sub systems.

Stand alone training simulators presently in use at INS Shivaji are like any other simulator of the type installed ashore the world over for operator training. These simulators utilize digital computers, connected to control consoles that are identical in operation and appearance to the ship's machinery control room . An instructor's console is connected to control the simulator and to introduce malfunctions, initialize the imulated operating states and guide the trainee during the conduct of the training. To make the training environment more realistic, the simulator includes various sound and light effects logically dependent upon the simulation situation.

SIMULATORS IN THE NAVY

The Indian Navy has already made tangible progress in the area of simulator training by introducing Tactical Trainers, Aircraft simulators, Navigation simulators, Submarine simulators, NBCD simulators, Power Generation and Propulsion system simulators. The first engineering simulator in the Navy was the MCR simulator for SNF class of ships. The simulator was designed and developed by BHEL, Hyderabad in collaboration with NSTL(V) at a cost of approximately Rs. 76 lakhs and was commissioned at INS Shivaji in May 1988. Thereafter an SNF Gas turbine starter trainer and also a 1241 RE Gas turbine starter trainer were made at Shivaji by in-house efforts, mainly through projects by trainee officers. Then, the 1241 RE MCR simulator was made by NSTL(V) and installed at the simulator complex in Shivaji. Recently, a M-36 Gas turbine starter trainet for the P-15 class of ships was also made at Shivaji as part of project by MESC officer. The project cost was approx. Rs 2.50 lakhs only.

Some of the areas where simulators can be effectively utilized in the Indian Navy are :-

(a) MCR Simulators
(b) Ship handling simulators
(c) Tactical trainers
(d) Aircraft/Helicopter simulators
(e) Weapon system/Fire control system simulators
(f) Maintenance training simulators for :

  (i) Missiles
  (ii) Radar
  (iii) Torpedoes
  (iv) Propulsion plants
  (v) Aircraft

(g) Submarines training simulators
(h) Nuclear propulsion plant simulators
(j) Communications/ EMC simulators
(k) PGD training simulators
(l) NBCD simulators

ADVANCES IN SIMULATION

Some of the hardware advancements leading to a revolution in training simulators are— high speed processing, networking, computing, versatile peripheral devices, graphics work stations and graphics user interfaces etc. Similarly, simulation software has now been equipped with highly efficient techniques, algorithms, tools and commercially available packages. Development of techniques in numerical methods and simulation languages for deriving discrete continuous models have made the creation of dynamic simulated replicas easily achievable. A large no. of PC based simulation packages (GPSS - General Purpose Simulation Software) are available in the market today.

With such exponential growth of the simulation aids in hardware and software, the exploitation of simulators for training is expected to extend its arms to every field of life. The intelligent tutorial system fast replacing the conventional simulator training method. This system automatically re-orients the training sessions according to the understanding abilities of the trainee, thus facilitating efficient utilization of the system.

Another futuristic trend in simulator hardware is to have multimedia interaction like operating sophisticated cameras, large screen displays, audio visual systems etc. simultaneously with the high speed processors.

CONCLUSION

The large number of advantages associated with the use of simulators for training have been reviewed. However, the awareness of the operator of being trained on a dummy equipment may at times effect his performance level. This may be offset by developing a performance evaluation schedule while the operator is trained on the simulator.

In view of the tremendous benefits associated with the use of simulators for training, it is quite essential for the Navy to promote use of simulators in every possible arena . It would be a useful idea to incorporate provision of a training simulator with every major equipment incorporated in the Navy.

No simulator can ever replicate complete realism. A training simulator is unlikely to reproduce the ambience of real life situations. Too much simulator exposure may tend to make operators insensitive to other sensitive inputs like sight, feeling, smell, sounds and vibrations, which in itself may lead to lack of response on the part of the operator. Good training must therefore strike a balance between time slots on simulators and live plants.

ENGINEERING PERSONNEL FOR MERCHANT NAVY

3

Sasadhar Datta
Fellow

ABSTRACT

Highly competent personnel, particularly at management level, are required for operating around forty thousand merchant ships all over the world. Emoluments being attractive persons from developing countries are eager to join the profession. Need for superior standard of qualification and experience of the marine officers is recognised internationally. Unfortunately, some countries do not hesitate to compromise with these vital requirements with the view of improving employment prospects of their nationals on board foreign ships, resulting in more number of accidents due to human failure. This ominous development has led the International Maritime Organization (IMO), an agency of the United Nations, to update provisions of the "Standards of Training, Certification and Watchkeeping (STCW) for Seafarers, 1978 by holding another conference in the same subject matters in 1995. In order to comply with the enhanced requirements, the Government of India has framed fresh examination rules for various categories of searfarers. This paper is an attempt to study whether the proposed changes will have the desired effect in raising training standards of Marine Engineers and turning them out in adequate numbers, not only to fulfill requirements of Indian merchant shipping but also for overseas employment. The latter will serve the dual purpose of widening technical experience and earning substantial foreign exchange for nation exchequer.

NUMBER OF MARINE ENGINEERS REQUIRED

A study made by the International Shipping Federation has predicted the following scenario pertaining to India at the end of this century:

Availability of Marine Engineers...9,000
Requirement of Marine Engineers..13,500
Shortfall..4,500
Number of Marine engineers to be trained annually to remove imbalance.....1,500
Percentage of Indian Marine Engineers employed on foreign ships....................50

Annual foreign exchange earning by this small community of 4,500 marine engineers employed overseas would be about rupees five hundred crores. Considering the vast technical potential of the country this amount can easily be doubled in a short period without any additional investment. A small country like Philippine, having population one fifteenth of India, has marine officers numbering five times that of this country. Even Indonesia is ahead of India in this regard.

EXISTING SOURCES

Currently, the following routes are available for joining the marine engineering profession:

Marine Engineering Research Institute
Graduation in Mechanical Engineering
Graduation in Electrical or Electrical & Electronic Engineering
Diploma in Mechanical Engineering
Engineering Apprenticeship after Higher Secondary (10+2) level Indian Navy
Sea-going Engine Drive certification

However, the number of persons trained annually is far short of requirements for Indian ships and overseas employment. The main cause of this deplorable situation is the stipulation of compulsory training in a marine workshop. Though national tonnage has increased manifold since independence, ship repairing and building within the country has neither grown nor modernised for providing requisite training facilities.

STANDARDS OF TRAINING, CERTIFICATION AND WATCHKEEPING (STCW) FOR SEAFARERS, 1978 AND 1995

In order to establish a uniform worldwide level of competence for seafarers the International Maritime Organization convened, for the first time, a conference of the member states in 1978 wherein regulations were drawn up pertaining to competence of various categories of personnel employed on ships. In course of time it became necessary to update the provisions of the Convention. Therefore, another Convention on the same subject was held in 1995 "to clarify the standards of competence required, introduce qualification requirements for trainers and assessors, provide effective mechanisms for enforcement of its provisions and allow greater flexibility in the assignment of functions on board ship and thus broaden the career opportunities of seafarers". Detailed model courses have also been prepared by the Organization for various subjects. An Alternative Training Scheme of thirty months duration, combining theoretical instruction ashore as well as shipboard experience, has been introduced for induction into marine engineering profession. This is a significant departure from existing practice followed in India.

EXAMINATION RULES, 1998

The Merchant Shipping (Examination of Engineer Officers in the merchant Navy) Rules, 1989 has been superseded by the Merchant Shipping (Standards of Training, Certification and Watchkeeping for Seafarers) Rules, 1998 with effect from 1st August 1998. The new rules are in conformation with the provisions of STCW as modified in 1995.

A major change is the Alternative Training Scheme of 30 months duration composed of the first 12 months at a shore institute, thereafter 9 months training as Cadet on board ships followed by the rest of 9 months again at the shore institute. Intake standard has been prescribed as "Senior Secondary School Certificate (10+2)". All the existing routes of entry have been retained and emphasis is continued to be laid on marine workshop training.

ALTERNATIVE TRAINING SCHEME (ATS)

As per the Model Course developed under the Norwegian Programme of the international Maritime Organization, during first phase of training lasting for twelve months the trainees are to devote :

483 hours for theoretical instruction in 9 specified subjects;
530 hours at workshop in 5 particular sections;
54 hours at laboratories.

Qualifications of entrants have not been clearly indicated. The minimum age for appearing in the initial Certificate Competency examination being 18 years, after the training period of 30 months and 6 months sea service, age of a trainee at entry stage works out to 15 years. In India a person of this age is not likely to be beyond Secondary (10) level. However, in the Examination Rules an entry requirement of ten plus two has been stipulated. Thereafter, the students are to undergo 9 months of training on boardship as Engineering Cadet. Last 9 months of the 30 months training are to be utilised for:

439 hours of theoretical instruction in 5 specified subjects;
300 hours at workshop for advanced practice and marine maintenance;
40 hours at laboratories.

Imparting the desired standard of knowledge and skills within the limited time frame would indeed be a formidable task. Lured by the possibility of making quick money a number of private entrepreneurs have come forward to set up the Scheme. Recent press advertisements of such organisations are indicative of the role that big money is going to play in training marine engineers. The Marine Engineering Research Institute (MERI) is, no doubt, best equipped to implement the Alternative Training Scheme without any investment and at moderate cost. Surprisingly, there is no move in that direction.

BRIDGING THE GAP

Even under the new Examination Rules, yearly availability of marine engineers in India is unlikely to exceed 500 against the annual demand of 1,500 thus leaving a chasm of 1,000. The main cause is continued insistence on prolonged training period in a marine workshop. It has been pointed out earlier that training capacity of marine workshops within the country has shrunk and quality of instruction has deteriorated. Taking advantage of the requirement under the Examination Rules, some of the marine workshops are extracting heavy amount, particularly from graduate apprentices, for imparting training without proper supervision, thereby making a mockery of the whole system. In view of the course content and associated exposure to workshop practice, Graduate Mechanical Engineers are, undoubtedly, most suitable for employment on ships. Unfortunately, though considered capable of taking up responsible positions ashore straight after graduation, they are compelled to go through one year workshop training of dubious utility before being allowed to join a ship in a junior position. They can certainly be engaged on ships as trainee marine engineers for six months immediately after obtaining degree. During this period the shipping company should arrange for their subsistence. For the next six months they may be employed as Assistant Engineer at a reasonable salary to enable them to make provision for attending three months compulsory preparatory course before appearing for the initial certificate of competency examination. With hundreds of engineering colleges in the country there is on dearth of graduate mechanical engineers. Every Indian and foreign ship employing marine engineers from the country can be persuaded to provide berth for one trainee and one assistant engineer from this stream. The scheme can be designed for an output of one thousand marine engineers annually to bridge the gap between supply and demand. The STCW Convention as well as the examination rules provide for adopting the measure.

THE EXAMINATION SYSTEM

Examinations for the certificates of competency are held every month, except in May, at the Mercantile Marine Departments, Calcutta, Chennai and Mumbai. However, the main bottleneck is critical shortage of examiners. Some countries have overcome the problem by availing services of outside agencies. Attendance of preparatory courses is now compulsory for appearing the examinations. Organizations providing such courses are required to assess the candidates as part of the Quality System monitored by independent certification agencies. Therefore, certificates of the approved organizations can be accepted and further written examination at the Mercantile Marine Department may be done away with. Approval should, however, be accorded to bodies of unimpeachable credentials like the Institution of Engineer (India) and the Institute of Marine Engineers (India). For conducting oral test a three member permanent board may be set up at each of the centres with one representative each from the Mercantile Marine Department, Institution of Engineers and Institute of Marine Engineers. As presence of all the three members cannot be assured, viva voce may be carried out by any two members. Such an arrangement will considerably ease the pressure on the Mercantile Marine Department and ensure sanctity of the examination.

CLOSURE

Needless obsession with quality, reluctance to change with time and lack of innovative thinking have led to perpetuation of serious imbalance between availability and requirement of marine engineer in India. Remedy lies in proper utilisation of existing resources rather than creating new facilities, which is expensive and time consuming. From these points of view, training of graduate mechanical engineers on board ships is most cost and time effective. Introduction of Alternative Training Scheme of shorter duration should not lead to the profession of marine engineering becoming an exclusive preserve of the wealthy. Difficulties arising out of recurring shortage of examiners can be overcome by availing services of learned societies like the Institution of Engineer (India) and Institute of Marine Engineers (India).

REFERENCES

1. International Shipping Federation—Final Report of the Indian Manpower Project.
2. International Maritime Organization—STCW Convention Resolutions of the 1995 Conference and STCW Code.
3. International Maritime Organization—Model Course 7.04.
4. Government of India—The Merchant Shipping (Standards of Training, Certification and Watchkeeping for Seafarers) Rules, 1998.
5. Government of India—M. S. Notice No: 11/98.

II

CRAFT VESSELS EQUIPMENTS AND INSTRUMENTATION:
Its design, construction, installation and operation

REMOTELY OPERATED VEHICLE FOR UNDERWATER APPLICATIONS

4

S. K. Ghosh, G. Purkayastha, S. Datta, R. Ghosh and S. N. Nandy
Central Mechanical Engineering Research Institute Durgapur-713209

ABSTRACT

ROV 200 MK-II is a remotely operated underwater vehicle designed and developed for carrying out various types of underwater jobs related to inspection and intervention upto a depth of 200 metres. The ROV is powered by five thrusters placed orthogonal directions to operate ROV along three Cartesian axes for optimum control of the system. The sensors for position and orientation feedback along with cameras and sonar are mounted on the ROV for proper monitoring of the vehicle. The system is controlled from a control console placed on the mother ship, where all the sensory informations are displayed on several display devices including video monitors. The control & telemetry of the system is based on ship board computer control or joystick control and high speed telemetry link. Here the surface computer controls all operator inputs through telemetry link and provides graphics display of various control, sensor and feedback data. Dynamic control of the system for path tracking and station keeping is done through PD filter using feedback from senors mounted on the ROV. It has a payload carrying capacity of 60 kg. The system has been rigorously tested for more than 500 hrs. in CMERI shallow basin as well as for more than 50 hrs. in open sea upto a sea state 2.

1. INTRODUCTION

As India becomes seriously involved in activities related to evaluation and subsequent exploitation of marine wealth and seabed mining, development of underwater systems for such future applications has gained national importance. Central Mechanical Engineering Research Institute (CMERI), Durgapur, is at present involved in design and development of a remotely operated underwater vehicle, ROV-200 MK-II, for operation upto a depth of 200 m. The ROV would become an essential part of the seabed mining system. It is also expected to have great potential for utilisation in other sectors, mostly in the Exclusive Economic Zone of India.

2. SYSTEM DESCRIPTION

ROV is an under-water robotics system, designed to carry different types of equipments, sensors and manipulators for carrying out various types of jobs related to inspection and intervention in highly unstructured sea environment.

Fig. 4.1 : The ROV being Launched for underwater testing

The mechanical system consists of a modularly designed open-frame structure with cylindrical buoyancy packages designed to withstand 200 metres of water column pressure and to provide a net positive buoyancy. This ensures that the system will come up to the surface in case of power failure, thus eliminating chances of losing it. Special sealing and pressure compensation techniques along with special materials are used to withstand saline corrosion and provide better mechanical properties.

The ROV is equipped with segmented landing pads for proper gripping and is powered by five 3 phase, 440V AC motorised propellers, also known as thrusters. Of these five thrusters, two are used for forward motion, two for the vertical motion and one for lateral motion. The rotational motion is achieved by running various combinations of forward and lateral thrusters.

The sub-sea control unit with pressure compensation device houses the control hardware for controlling thrusters and the sensors. The bulkhead type modular underwater connectors are used with the control unit for quick assembly and dismantling of the cables.

The launching system consists of a depressor, adjusting weight, pulley, motorised winch and a hybrid slip-ring system. The motorised winch in conjunction with the slip-ring system is used to facilitate the easy handling of the 400 m long cable having a diameter of 45 mm in the untwisted condition.

The system is also equipped with a high resolution underwater camera and navigational sensors like acoustic positioning sensor, sonar, 6-axes gyro, and heading, tilt angle and depth sensors. These sensors provide feedback from the ROV for implementing dynamic position control of ROV.

The ROV is controlled from a surface control unit placed on the mother ship, where all the sensory information are displayed on several display devices, including video monitors. The surface computer controls all operator inputs through telemetry link and provides graphics display of various control, sensor and feedback data. Braking feature is provided for instant stopping of ROV during its motion along a predefined path and in case of exigency, a manual joystick mode is provided which overrides computer controlled mode during its motion. The path tracking data is recorded in a separate file for future analysis.

3. DYNAMIC MOTION CONTROL

Underwater dynamic motion control has become a major area of research for such important applications as inspection and monitoring of various underwater systems, for sea-bed mining and for performing various complicated underwater tasks. The software for dynamic motion control of ROV consist of

(i) path-tracking algorithm
(ii) station-keeping algorithm

The path-tracking and station-keeping algorithm is guided by various feedbacks at precise time intervals, from the sensors on the vehicle. Any deviation from the defined path and the defined heading is corrected by using PD control described below.

The ROV motion is that of an unconstrained rigid body i.e. it possesses six degrees of freedom, viz. three translational degrees and three rotational degrees. The three translational coordinates are expressed as three dimensional vectors in an intertial frame and the other three components are generally expressed by three eulerian angles - roll, pitch and yaw.

In the present case, roll and pitch are maintained within the acceptable limits through proper designing and positioning of the buoyancy pack. The following exercise is therefore limited to the control of three translational degrees and one rotational through different configurations of thrusters. This enables the ROV to move along a predefined path.

In its full complexity, the task of controlling an ROV is indeed difficult. The problem is thus generally broken into some sub-problems which are reasonably decoupled. As the roll and pitch are within the acceptable limits, the horizontal and lateral trusters remain in the horizontal plane and the vertical thrusters remain in the vertical plane. This allows the control in the vertical direction to be decoupled from the control in the horizontal plane. The decoupled scheme for low speed ROVs moving either vertically or horizontally at a time consists of

(a) heading correction
(b) depth control
(c) planar path control

These are implemented through some forms of linear feedback controller and in this case, PD scheme is used.

3.1 Path Tracking Algorithm

The input is provided in terms of thc target to set the desired path of motion that does not take into account any external disturbances. Such a situation, where there will be no external disturbances, is hypothetical, and the ROV is likely to be off the planned path due to sea drift.

For implementing the path-tracking algorithm, the external force due to sea drift is considered. The drifting force is represented by two components, namely,

(a) along the direction of path,
(b) along the path orthogonal to it.

Formulation for simultaneous path tracking with heading correction is achieved by negating external forces or moments acting on the system, which is described below.

The total control of path tracking algorithm is divided into three modules. Module 1 describes heading correction, taking into account torque generated due to sea drift, unbalanced drag force and damping torque generated due to rotational motion of ROV. The resultant torque Γ_R along the z-axis is given by

$$\Gamma_R = \Gamma_i - \Gamma_D - \Gamma_d \quad (1)$$

Where

Γ_R = resultant torque generated

Γ_i = torque generated due to current sent to forward tnrusters using PD scheme

Γ_D = damping torque acting against the rotational motion of ROV

Γ_d = torque generated due to sea drift

$$\Gamma_i = T_{RI} * a_R + (-T_{LI}) *(-a_L) \quad (2)$$

where

T_{RI} = thrust generated by right thruster for heading correction

T_{LI} = thrust generated by left thruster for heading correction

a_L = left moment arm from centre of drag plane to left thruster

a_R = right moment arm from centre of drag plane to right thruster

The drag plane in this case is a plane passing through the centre of drag of the frontal face and is perpendicular to the line joining the two forward thrusters.

Module 2 describes the orthogonal deviating error correction due to the component of sea drift acting perpendicular to the path of motion. Hence, the resultant force acting on the body necessary for correcting this error is given by

$$F_{RLat} = F_{iLat} - F_{DLat} - F_{dLat} \quad (3)$$

Where

F_{RLat} = resultant force generated in lateral direction

F_{iLat} = force generated due to current sent to thrusters using PD scheme for lateral correction

F_{DLat} = damping force generated due to translational motion of ROV

F_{dLat} = force generated due to sea drift in lateral direction

$$F_{iLat} = \sqrt{(T_{R2} + T_{L2})^2 + L_2^2} \quad (4)$$

Where

T_{R2} = thrust generated by right thruster for lateral correction

T_{L2} = thrust generated by left thruster for lateral correction ($=T_{R2}$)

L_2 = thrust generated by lateral thruster for lateral correction

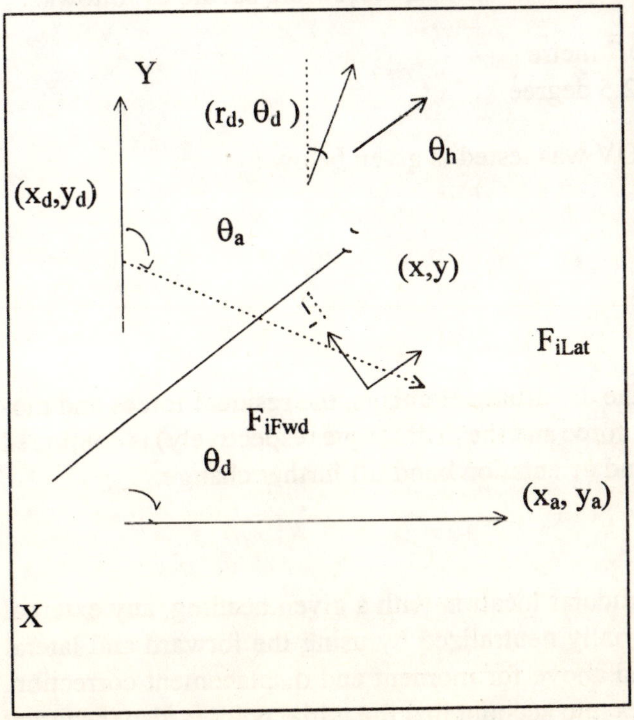

Module 3 describes the force acting on the body due to component of sea drift acting in the direction of motion. Hence, the resultant force acting on the body is given by

$$F_{rfwd} = F_{ifwd} - F_{dfwd} - F_{dfwd} \quad (5)$$

F_{RFwd} = resultant force generated in the direction of motion

F_{iFwd} = force generated due to current sent to thrusters using PD scheme for motion along the defined path

F_{DFwd} = damping force generated due to translational motion of ROV

F_{dFwd} = force generated due to sea drift in direction of motion

$$F_{ifwd} = \sqrt{[(T_{R3} + T_{L3})^2 + L_3^{\,2}]} \quad (6)$$

where

T_{R3} = thrust generated by right thruster for motion along prescribed path

T_{L3} = thrust generated by left thruster for motion along prescribed path (=TR3)

L_3 = thrust generated by lateral thruster for motion along prescribed path

Hence final thruster output is given by vectorially adding all three modules. Therefore,

$$T_R = T_{R1} + T_{R2} + T_{R3} \leq T_{Rmax} \quad (7)$$

$$T_L = T_{L1} + T_{L2} + T_{L3} \leq T_{Lmax} \quad (8)$$

$$L = L_2 + L_3 \leq T_{max}$$

where T_{Rmax} and T_{Lmax} are maximum thrusts that can be exerted by each thruster.

This algorithm has been tested on the ROV developed at CMERI both in controlled conditions and in open sea. The boundary conditions those were set are as follows:

1. Path bandwidth -- +0.7 metre
2. Heading bandwidth -- + 2.5 degree

The sea conditions in which the ROV was tested is given below :

1. Wind speed - 14 knots
2. Surface current - 1.2 kmots
3. Wave height - 1.5 metres
4. Depth of testing - 40 metres

To render the system proactive to the disturbing elements, the residual forces and moments (equal to the orthogonal drift force and the drift torque respectively) is continued even after entering the path band and orientation band till further change.

3.2 Station Keeping Algorithm

Once the ROV is anchored at a particular location with a given heading, any external forces acting on the ROV is vectorially neutralized by using the forward and lateral thruster using the scheme described above for moment and displacement correction. The station-keeping algorithm takes into account that the initial point is also the target point. As before, the different components of the forward and the lateral thrusters are used for compensating the deviations in the orthogonal, parallel and the heading parameters.

4. FUTURE SCOPE

The sea-bed mining programme is targeted for a depth of 6000 metres. The future ROVs, will have to operate at this depth. These will be used in tandom with the mining system for supervision and monitoring. Those ROVs can also be used to more fully exploit the resources of the sea and also to locate and intercept targeted objects. Other areas of application may include BOP equipment recovery, visual inspection of sea-bed, crack detection of selected underwater joints, mapping and photo documentation of marine growth, actuation of tree connector overrides, etc.

The services of ROVs can be utilised by the ONGC, IOC, GSI, Navy, Port Authority, NIOT, NIO and other organisations related to offshore activities. Apart from these, the service of the ROVs can be used by the Indian Ocean rim countries.

5. CONCLUSION

The ROV has already been tested for more than 500 hrs. in the shallow basin facility of CMERI for evaluation of its performance. During this phase of the trial testing, it has successfully conformed to the requisite criteria for high static and dynamic stability, high maneuverability, straight forward motion, straight reverse motion, left turn, right turn, centre turn, hovering at a particular depth and floating in case of power shut down. After this, the ROV was trial tested first at the Chhenai harbour and then at open sea for ascertaining the seaworthiness of the system. The test was carried out in presence of the National Institute of Ocean Technology, Madras personnel from 'Sagar Purvi' a ship belonging to the Department of Ocean Development, New Delhi. The system demonstrated its under-sea inspection capability by inspecting the ship's propeller using an under-water camera and sonar from a distance of 8 m.
The test has substantiated the following technical achievements :

1. Sea worthiness of the total system.
2. Working of the sensors.
3. Computer controlled mode of ROV operation.
4. Maintaining a depth of upto a level of 40 mtrs.

In view of the successful testing of the systems at sea and a CMERI, it can be summarised that the following capabilities have been achieved regarding underwater technologies for ROV:

1. Structural design technology and fabrication of underwater systems.
2. Underwater sealing.
3. Control system design and implementation of control algorithm for mobile underwater system with 6 degrees of freedom.
4. GUI driven man machine interface software for handling multiple sensors.
5. Worthiness of the system upto sea state 2.

RISER SYSTEM FOR DEEP SEA BED MINING USING U TUBE CONCEPT

5

S. K. Ghosh
Scientist, Central Mechanical Engineering Research Institute
Durgapur - 713209

ABSTRACT

A scheme for the riser system of a sea-bed mining system has been formulated here. The concept uses the advantages of a U tube, where the levels of liquid in the arms are at the same level when both the ends are open, irrespective of the diameter variation of the tube. The scheme describes a total system for lifting solid particles through fluidisation, including the pumping system, shipboard collection system, feeding of the solid particles to the tube, and dynamic controlling of the tube profile. The scheme uses mud pump on shipboard to pump in sea water to one leg of the U tube and a screw pump to feed concentrated solution of the solid particles and water to the riser pipe. The flow in the upcoming pipe is enhanced by using the leg as input to the suction pipe of the mud pump. Using the U tube system also increases the rigidity of the profile. Mathematical formulation for a particular case has also been developed for the system. This system is expected to be useful for shallow-depth offshore sand mining and also for deep sea-bed mining.

Symbols used are as follows :

g	=	Acceleration due to gravity
γ_s	=	Specific gravity of solid particle
γ_w	=	Specific gravity of water
D	=	Nominal diameter of pipe
d	=	Diameter of solid particle
C_s	=	Concentration of solid particle
q	=	Spatial concentration of solids
L	=	Length of pipe
i_w	=	Pressure gradient for water
i	=	Pressure gradient for mixture
C_d	=	Co-efficient of drag for solid particle

λ	=	Co-efficient of friction factor for water flow in pipe
V_{to}	=	Terminal Velocity for a single solid particle
V_t	=	Terminal Velocity when solid concentration is Cs
V	=	Mean velocity of the mixture
V_f	=	Velocity in the feeder pipe
M_f	=	Mass flow rate in the feeder pipe
M_w	=	Mass flow rate in the down coming pipe of the U tube
V_i	=	Velocity of water from down coming pipe of the U tube at jetting point.

1. INTRODUCTION

The concept of deep sea-bed mining has been around for the last few decades. Some of the countries have already developed some prototypes for the same. Technological advances and explorations are taking place in these countries to incorporate new ideas for better mining, though efforts for commercial exploitation have been suspended for the moment.

India is a Pioneer Investor nation and has an area of around 7.5 lakh square kilometers earmarked in the Central Indian Ocean basin for future explorations and exploitation. India is also under the obligation of developing a viable deep sea-bed mining system. To pursue this, the Department of Ocean Development had in 1989 entrusted to the Central Mechanical Engineering Research Institute, Durgapur, India the task of developing a suitable deep sea-bed mining system. CMERI initially developed a land-based system for mining which was subsequently tested in shallow water. The Riser system responsible for transportation of the solid polymetallic nodules to the mother ship, initially comprised a bucket-chain system. Later, trials took place with a pneumatic aided hydrodynamic transport system. The pilot system was of 10 metres height. Subsequent development of the whole system is taking place at the National Institute of Ocean Technology, Chennai, India.

In this paper, a concept has been described which uses the advantages of a U tube for designing the riser system. The scheme describes a total system for lifting solid particles through fluidisation, including the pumping system, shipboard collection system, feeding of the solid particles to the tube, and dynamic controlling of the tube profile.

2. DESCRIPTION

The system for the lifting of polymetallic nodules which are mined from the seabed consists of a U-tube hung from the mother ship. The U-tube is hung from a universal joint to allow the swaying of the tubes due to relative movement of the the ship and the water mass. The U-tube has inter-spacing support structures added to it for preventing the two legs from getting tangled up and maintaining the profile of the legs. To maintain the profile of the tube a set of two thrusters at orthogonal directions, a gyroscope and an accoustic position detector are positioned at specific intervals. The thrusters are actuated through computer programme that utilises feedback data from the sensors for station keeping. This ensures that there is no substantial deflection of the tube. In the riser part of the tube there are interspaced valves which open during operation and close when the flow stops. This ensures that there is no clogging in case of system failure. Water is constantly fed by suitable pumps (preferably mud pumps) to one end of the U-tube. Standby pumps may be kept ready in case of emergency. The delivery pipe is inserted into a separator. The separator may be of the cyclone type, sedimentary type, etc. The suction pipe of the pumps starts from the top of the separator. In this way it is always possible to flood the pump positively on the suction side. The polymetallic nodule feeder tube is connected to the bottom of the rising part of the U-tube as shown in the figure. These nodules are fed in the form of slurry at a pressure higher than the pressure at the bottom point. The feeder consists of sets of screw type pump and reciprocating pumps (preferably pacemaker pumps of cementing unit) placed at the bottom of the screening system and the crusher of the mining crawler. The entry point has a unidirectional valve which ensures that the backpressure does not affect the feeding of the nodules. After being fed into the rising part of the tube, nodules get carried up by the momentum of

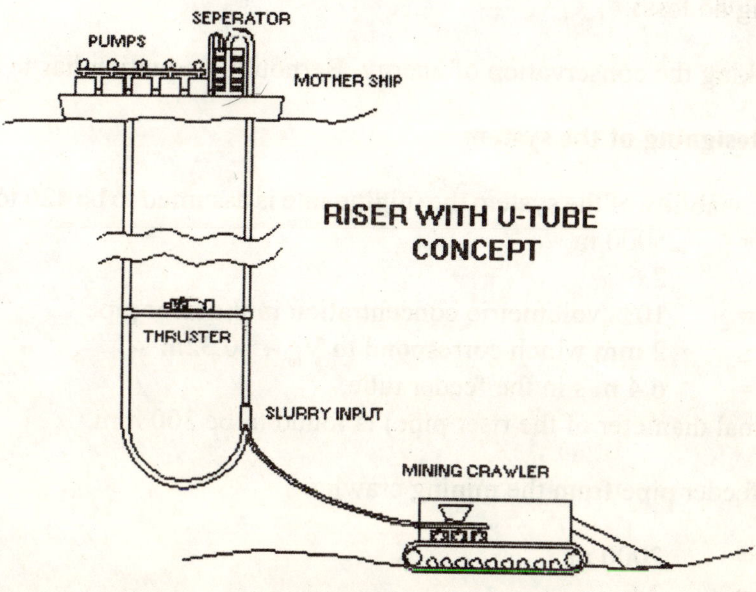

water. For better flow through the riser pipe, water jetting or air injection may be done at suitable points.

3. THEORY

For the upward transport of solid in a liquid medium, the important parameters are the velocity of the water through the vertical pipe, medium density, medium viscosity, coefficient of drag, size of the solid, density of the solid, pipe diameter, wall factor and concentration of solids. The formula provided below is for the settling velocity of solids in an infinite medium with solid concentration Cs.

$$Vt = [4 * g * (\gamma s - \gamma w) * d / (3 * \gamma w * Cd)]^{1/2} * [1 - (d/D)^2] * e^{-2.6 * Cs} \quad (1)$$

The static pressure gradient in the pipe carrying the solids,

$$iw = L * g * Cs * (\gamma s - \gamma w) + (1 + \gamma * L/D)/2 \quad (2)$$

The total pressure gradient in excess of the static head

$$i = i_w * (V_w/V)2 + q * (\gamma_s - \gamma_w)/ g_w \quad (3)$$

The final momentum of the entrained slurry and the initial mass of water must be provided by the momentum of the driving mass of water and the momentum of the incoming slurry mass from the feeder pipe.

$$\text{i.e. } M_w * V_i + M_f * V_f = (M_f + M_w) * V \quad (4)$$

(assuming no loss)

For checking the conservation of energy, Bernoulli's equation has to be applied.

3.1 Sample designing of the system

For economic viability of the system the mining rate is assumed to be 420 tons per hour.

L = 5000 m
γ_s = 2.6
C_s = 10% volumetric concentration in the riser pipe
d ≤ 2 mm which correspond to V_{to} = 0.32m/s
V = 6.4 m/s in the feeder tube,
D (nominal diameter of the riser pipe) is found to be 300 mm.

For the feeder pipe from the mining crawler

L = 2000 m
γ_s = 2.6
C_s = 40% volumetric concentration in the riser pipe
d ≤ 2 mm which correspond to V_{to} = 0.32m/s

V = 6.4 m/s in the feeder tube,
D (nominal diameter of the feeder pipe) is found to be 150 mm.

For the driving (down coming) pipe

L = 5000 m
V = 3.1 m/s in the feeder tube,
Amount of water handled = 1212 M^3/hr
D (nominal diameter of the driver pipe) is found to be 375 mm.
Jet velocity Vi is 7.2 m/s and the jet diameter is 240 mm.

Considering the effect of the solid along with the water, the static pressure at the bottom of the riser tube will be 80 kg/cm^2. The head loss due to friction in the riser tube is approximately 15m per 100 m and in the down coming tube, this is approximately 3 m per 100 m. So the total loss is 900 metres i.e. 90 kg/cm^2. The necessary pressure (after taking into account the head for imparting velocity) to be generated by the on-board pumps is 175 kg/cm^2. The head loss in the feeder pipe due to friction is approximately 20m per 100m, so the total head loss is (taking pipe length to be 2000m) 400m i.e. 40 kg/cm^2. So the pressure to be generated by the crawler board pumps is 122 kg/cm^2. Together with this air can also be injected at suitable points in the riser tube which will not only accelerate the flow but also reduce the pressure on the down coming pipe and the riser pipe.

For maintaining the profile of the U-tube, a station keeping algorithm is created. The mother computer, after obtaining inputs from the gyrometer and the accoustic position sensor, sends the necessary control signals to the thrusters to minimise the deflection. The control works on PID principle and is a close loop system.

The onboard pumping system consists of five mud pumps of 1700 hp each connected in parallel. The operating pressure of each of the pump is kept at 180 to 200 bars. The pumping capacity of the five pumps is kept at 1250 M^3/hr.

The mining crawler board pumping systems consists of three sets of pacemaker pumps and screw pumps. The screw pumps are used as feeder to the pacemaker pumps. The sets are connected in parallel. The operating pressure of each pacemaker pump is kept at 140 to 160 bars. The pumping capacity is maintained at 432 M^3/hr. The pulsation dampers are suitably designed for the operation.

CONCLUSION

The system provides a clog free and continuous lifting of solids, lighter mining crawler, better servicing facility and better profile maintenance of the riser tube. The above system can also be successfully used for lifting of the special sands from the shallow coastal depth by IRE.

DESIGN CRITERIA OF PIPING SYSTEM FOR SHIPS

6

A. Roy Chaudhuri and U.P. Thakur
DMD Establishment, Secunderabad

ABSTRACT

Proper functioning of ship machinery, operating under severe environmental conditions, depends to a large extent on adequate and correct design of inter-connected piping systems. Although, flexibility analysis of pipelines is fairly well established and computerised, the overall design methods still depend mainly on the experience of the designer. Unlike other equipment, finer aspects of pipework design are not clearly spelt out in text books/literature and depend largely on practical aspects.

This paper presents a practical approach for carrying out the piping flexibility analysis. It also describes other design aspects like dynamic loading and fluid hammer effect on the piping system and various remedial measures to be taken at the design stage.

INTRODUCTION

Many efforts have gone in the field of piping design and flexibility analysis in the last few decades. Various researchers have attempted to solve the problem of flexibility from different angles. Some of the prominent works are listed. While the method by Owens[6] is fundamental in nature, those[3-5] have been major contributors to piping flexibility analysis for many years. Advent of computer technology in engineering applications has led to the development of matrix methods in piping flexibility analysis. The computerisation of piping flexibility analysis is mainly attributed to Chen[9] who extended the early method of Brock[1] and presented a valuable and easily programmable method based on stiffness matrix. Natarajan[7] proposed a simplified method of flexibility analysis based on stiffness method, which is particularly useful for simple piping systems. A comprehensive matrix method in piping flexibility analysis incorporating different intermediate partial constraints was proposed by Kazem Abhary[8]. This method integrates intermediate partial constraints into the body of the equilibrium equation of piping system.

In addition to the requirement of piping flexibility, consideration of piping transients, which is increasingly becoming a major factor in system design process for marine propulsion, have gained significant importance. It is necessary to assure the integrity of piping systems under transient condition. The quick manoeuvring of ships require that the load changes in the main propulsion turbines be very fast. This design criterion has dictated the development of very quick acting turbine stop valves and closing time of such valves is as small as 0.1 sec. Steam hammer, which originates from sudden closure of turbine stop valves, has been known to produce relatively large-scale vibration in main steam lines. The measures that can reduce the effects of this steam hammer should be taken during the design stage itself in order to eliminate time consuming costly shut down for modification of pipe supports.

This paper presents a general approach to piping flexibility analysis and practical design aspects like usage of restraints, anchors and cold pull up etc., particularly for constrained piping layout in naval applications. Also, it highlights the need of considering the piping transients like fluid hammer effect on the piping system and various remedial measures that need to be taken at the design stage.

2. PIPING FLEXIBILITY

As the theoretical base of the piping flexibility analysis is fairly well established, commercial software are available for the flexibility analysis. However, still the actual goal of the flexibility analysis and piping design largely depend on the experience and technical judgement of the designers. In order to achieve an optimum piping layout, the designer has to exercise various options like directional restraints, partial constraints, anchors and cold springs etc. Therefore, these options are equally important compared to the solutions of the flexibility equations as it directly affects the performance, cost and manufacturing aspects.

2.1 Piping Design for Marine Steam Turbine Plants

The piping layout and design of marine steam turbine plants differ from the shore based installations. In power plant or oil refineries, the space is not a genuine constraint for piping layout and, therefore, flexibility problems are largely solved by incorporating loops in the piping systems. However, in ship installations, the layout is dictated by very small compartments with large number of equipment confined in that space. Therefore, it is physically not possible to have sufficient number of loops in the piping systems in order to achieve the flexibility and it rests on the designer how he solves the problem of flexibility.

As the possibility of providing sufficient number of expansion loops is remote, the designer has to study the problem more closely in order to exercise other options like introducing restraints, anchors and cold springs. While each of these options may reduce the stress problem in a particular zone of the piping systems, the other zones may adversely be affected in terms of stress or terminal reaction. It has been observed in many instances that by putting directional restraints in a particular area, while expansion

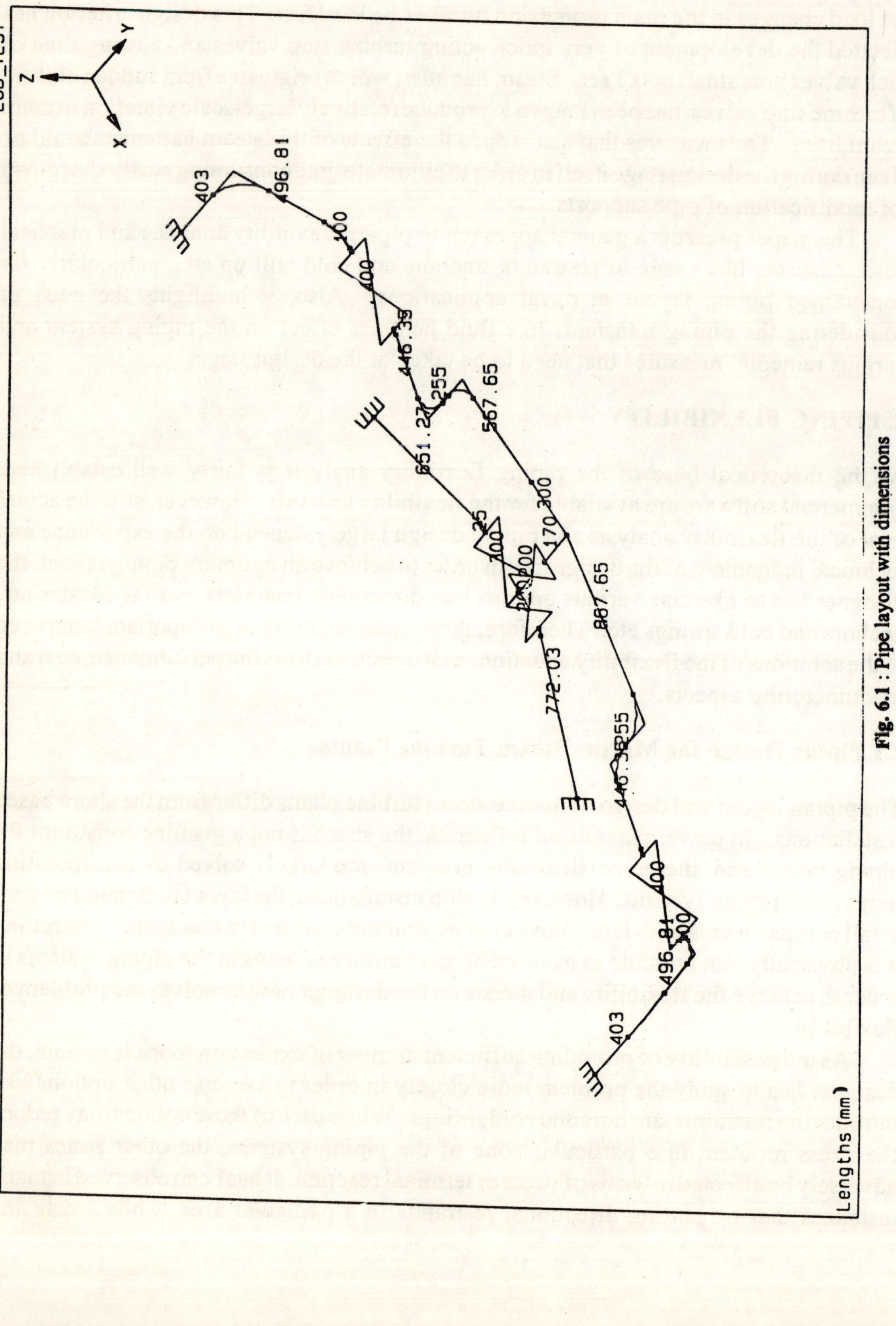

Fig. 6.1 : Pipe layout with dimensions

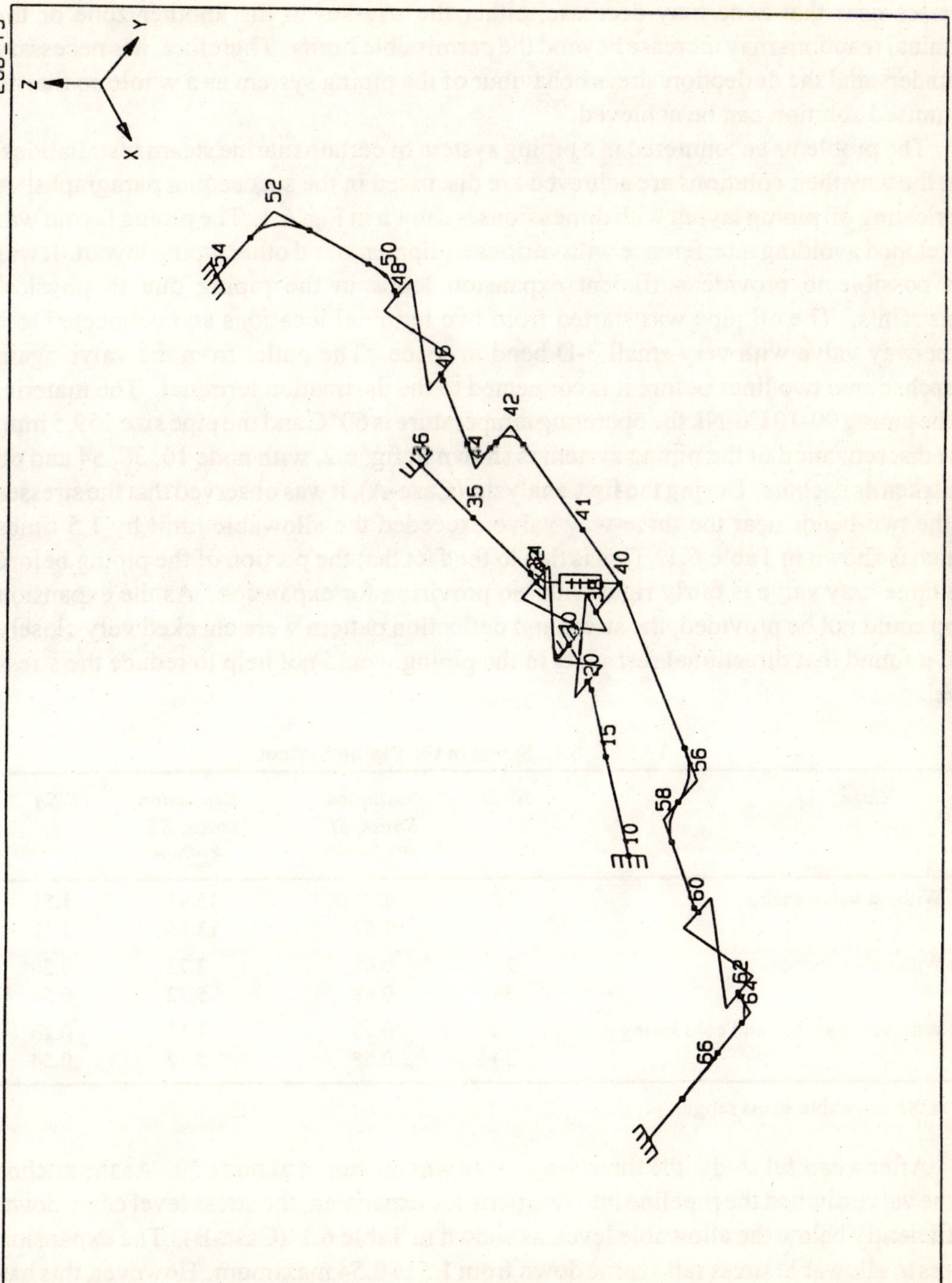

Figure 6.2 : Discretisation of Piping

stresses near that zone may decrease, either the stresses in the another zone or the terminal reactions may increase beyond the permissible limits. Therefore, it is necessary to understand the deflection/stress behaviour of the piping system as a whole so that an optimised solution can be achieved.

The problems encountered in a piping system of certain marine steam installations, and the way their solutions are achieved are discussed in the succeeding paragraphs. A lubricating oil piping layout with dimensions is shown in Fig. 6.1. The piping layout was developed avoiding interference with various equipments and other piping layout. It was not possible to provide sufficient expansion loops in the piping due to physical constraints. The oil pipe was started from two terminal locations and connected to a three-way valve with very small 3-D bend in space. The outlet from the valve again branched into two lines before it is connected to the destination terminal. The material of the piping 90-10 Cu-Ni, the operating temperature is 60 °C and the pipe size 159.5 mm. The discretization of the piping system is shown in fig. 6.2. with node 10, 36, 54 and 68 are taken as anchors. During the first analysis, (Case-A), it was observed that the stresses in the two bends near the three-way valve exceeded the allowable limit by 1.5 times which is shown in Table 6.1. This is due to the fact that the portion of the piping before the three-way valve is fairly rigid with no provision for expansion. As the expansion loop could not be provided, the stress and deflection pattern were checked very closely and it found that directional restraints in the piping would not help to reduce the stress level.

TABLE 6.1: **Stress in the Piping System**

Case	*Node*	*Sustained Stress, SL (Kgf/mm²)*	*Expansion stress, SE (Kgf/mm²)*	*SE/SA*
(A) Without valve anchor	20	1.71	15.91	1.51
	34	1.52	13.85	1.32
(B) With valve anchor	20	0.68	2.73	0.26
	34	0.68	5.72	0.54
(C) With valve anchor and cold spring	20	0.68	2.13	0.26
	34	0.68	5.72	0.54

SA is the allowable stress range.

After a careful study, the three-way valve was anchored at node 30. As the anchor in the valve divided the pipeline into two parts for expansion, the stress level came down sufficiently below the allowable level, as shown in Table 6.1 (Case-B). The expansion stress to allowable stress ratio came down from 1.5 to 0.54 maximum. However, this has created another problem as the terminal force at anchors 10 and 36 increased drastically to 12847 kgf and 13477 kgf respectively as shown in Table 6.2.

TABLE 6.2: **Forces at the Terminal**

Case	*Node*	*X-Direction force (kgf)*
(A) Without valve anchor	10	3880
	36	–3840
(B) With anchor at valve	10	12847
	36	–13477
(C) With valve anchor and cold spring	10	7460
	36	–7817

When all other options to reduce these forces failed, the option of cold springs was used at node 15 and 35. Cold springs are used to pre-stress the piping so that the terminal force reduces during operating conditions. The result of using the cold spring is shown in Table-6.2 (Case-C) where it reduced to 7817 kgf. In this way an optimum solution was achieved where stresses and the forces were within acceptable limits. Even though, the solution looked simple, it required the designer's practical experience and knowledge about structural mechanics of the piping system. This is a small example of the nature of flexibility problems encountered in piping system design of ships. In this way, the designer has to consider every individual piping system and the solution of each system is unique by itself.

3. PIPING TRANSIENTS

In theory, instant closing of a valve in the pipeline will burst the piping, assuming that the flow can be stopped in zero time and the fluid and pipe walls are wholly elastic. However, in practice all valves have finite closing time. The design of marine steam plants requires that the load changes in the main propulsion turbine should be fast and the turbine protection system along with the main steam stop valve should close quickly in case of sudden load throw-off due to gear coupling failure. The closing time of the stop valve is decided based on the basis of permissible overspeed of the turbine during the load throw-off. This situation has led to the development of stop valves whose closing time is approximately 100 milli-sec or even less. If adequate measures are not taken in the design of the piping system and its auxiliaries, catastrophic failure may occur.

3.1 Theory of fluid hammer

During the closing cycle of a valve, the kinetic energy of the moving fluid is converted into potential energy. Elasticity of fluid and pipe wall produces waves of positive and negative pressure which propagate with sonic velocity. These pressure waves travel from the valve through the fluid to the steam generator and back to the valve. This process would keep repeating except that internal friction and other energy losses gradually damp out the waves.

The initial shock and suddenly stopped flow can induce transient pressure changes that exceed the static pressure. The time T_0 of one round trip of the pressure wave from valve to reservoir/steam generator and back is the critical time that establishes the severity of the fluid hammer. If the valve closes within the critical time, the pressure rise is the maximum. It does not matter how much faster than T_0 the closing time is as long as fluid stopped completely before the first pressure wave completes the round trip. If the valve does not close within critical time T_0, the maximum pressure rise will be comparativelv less.

3.2 Design remedies against dynamic loading

Given certain steam conditions and initial velocity, the pressure surge upstream of the suddenly closed valve would depend on the closing characteristics of the valve. Each valve's closing characteristics produces a pressure surge pattern unique to those specific characteristics[10]. Because the dynamic loading on a piping system is the result of these pressure surges, it is concluded that there must be one specific pressure surge pattern that will result in the lowest magnitude of dynamic loading. Mikasinovic, et al.,[11] had developed a procedure to find the optimum valve characteristics for various initial flow velocities which would produce minimum dynamic loading on the piping system. His work is based on the principle that linear pressure surge will result in the lowest dynamic loading on piping.

Therefore, at the design stage, the designer can work out the flow velocities in the piping system based on the mass flow rate and pipe size. The optimum closing characteristics of the main steam stop valve can be decided by the designer before the placement of the order which will produce minimum dynamic forces in the piping system. Further Mikasnovic[11] et. al. are several graphs to workout the amount of dynamic forces.

However, in practice, it is difficult to secure a turbine stop valve with the required characteristics of the user. Generally, the valve manufacturers follow their standard design for these costly valves. Therefore difficulties normally arise to get a suitable valve for a particular vessel's requirement. It is, however, possible to overcome this problem by incorporating additional rigidity in the piping system so that dynamic loading will not lead to any failure.

3.3 Piping design considerations for dynamic loading

Some practical considerations at the design stage may reduce piping transient problems in a big way. From the dynamic loading point of view, the piping, supports and spring hangers should be as rigid as possible. However, from the thermal and flexibility analysis point of view, the piping system should have enough flexibility so that the expansion stresses induced in the piping system are less than the allowable value. These two requirements are completely opposite to each other. Therefore, a compromise has to be made while carrying out the flexibility and dynamic analysis. The most critical areas of failures under dynamic loading are spring hangers. Normally, spring hangers are selected on the basis of piping load at that point and the maximum vertical deflection arising out from the expansion of the piping. This procedure may lead to the failure of the spring hanger under dynamic loading as deflection during dynamic loading may exceed the allowable range of the spring. Therefore, the designer should consider the vertical deflection due to expansion as well as dynamic loading together and select the spring hanger.

Another aspect of spring hanger selection for dynamic loading is to increase the rigidity. Variable spring hangers are likely to give more rigidity than the constant spring hangers even though in some cases, the constant springs would be preferable from flexibility point of view. Therefore, as far as possible, more variable spring hangers should be used in the piping system where dynamic loading are significant. Directional restraints and anchors in the piping system increase the rigidity and should therefore be used as much as possible keeping the expansion stress within the limit. However, limit stops/partial restraints are not advisable from dynamic loading point of view.

3.4 Effect of transients on flexibility

It is seen that the goal of dynamic analysis is to increase the rigidity of piping system so that it does not fail under transient dynamic loading. However, in order to achieve this, the piping system loses flexibility and the expansion stress may exceed the allowable limit. Therefore, the dynamic analysis and flexibility analysis have to be carried out simultaneously and any changes made during the dynamic analysis be reflected into the flexibility analysis. If it does not meet the flexibility requirement, changes may be required in the restraints, anchors etc. and dynamic analysis is to be rechecked. This way, few iterations of dynamic and flexibility analyses are required to be carried out to achieve a finally converged solution.

4. CONCLUSION

Piping design and analysis of marine steam plants are fairly complex, as design requirements are many-fold and sometimes contradictory to each other, calling for both flexibility and rigidity in design. This paper has dealt with the practical aspects of carrying out the piping flexibility analysis which are normally not available in the text books. It has also described the design requirements for piping transient/fluid hammer and the remedial measures to be taken at the design stage. It stresses the need of carrying out the flexibility and dynamic analyses simultaneously for piping systems design of ships in order to ensure structural integrity of the system.

REFERENCES

1. Brock, J.E., A matrix method for flexibility analysis of piping systems, Journal of Applied Mechanics, 19(4) (Dec. 1952) pp. 501-16.
2. Markl, A.R.C, Piping flexibility analysis, ASME Trans., 77 (2) (Feb. 1955), pp. 127-43.
3. M.W. Kellog company, Design of piping systems, New York, 1st edition 1941, 2nd edition 1956.
4. Peck, L.G., Meyer, R.F., Strong, P.F. and Kalson, H., The automatic calculation of forces and deflections in piping system, Transactions ASME, 80(1) (Jan. 1958), pp. 235-44.
5. Brock, J.E., Piping Handbook, 5th edition [King, R.C., (Ed)], chap. 4, McGraw Hill, New York, 1967.
6. Owens, R.H., Elementary development of piping flexibility analysis, American Society of Naval Engineers Journal, 72 (1) (Feb. 1960), pp. 159-70.
7. Natarajan, R. A Simplified pipe flexibility analysis programme—stiffness method, computers and structures, 22(3) (1986), pp. 299-305.
8. Kazem Abhary, Inclusion of Intermediate partial constraints in piping flexibility analysis—A general approach, Internal Journal of Pressure Vessels & Piping, 31 (1988), pp. 43-54.
9. Chen, L.H., Piping flexibility analysis by stiffness matrix, ASME Transactions, Journal of Applied Mechanics, 26 (4) (Dec. 1959), pp. 608-12.
10. Mikasinovic, M., Tung, P.C., Predict maximum steam hammer force with quick reference - chart method, Power (June 1981), pp. 63-65.
11. Mikasinovic, P.E., Ron Hoy, P.E., Good valve choice cuts steam hammer, Power (May, 1982), pp. 48-50.

CRAFTS/VESSELS/EQUIPMENTS RELATED TO "OCEAN MINING"

7

Lt. G. Ravi
CBIU, Mumbai

INTRODUCTION

The process of recovering mineral wealth from seawater and from deposits on and under the sea floor is receiving increasing attention. The reasons for this sudden emphasis are, the demand for mineral commodities is increasing at an alarming rate, and most of the developed countries have thoroughly explored for surface outcroppings, advances in exploration and exploitation of low grade and unconventional sources of ore, corresponding advances in oceanology have highlighted the importance of the ocean as a source of minerals and indicated that the technology required for their exploitation is in some cases already available.

The extended jurisdiction concept which divides the oceans into three separate jurisdictional zones; Territorial, Contiguous and Exclusive Economic zones, has come into effect on 16 Nov. 1994 according every nation the freedom to explore and exploit the resources of the ocean. It has also contributed to the increased awareness.

ORIGIN OF OFF-SHORE MINERAL DEPOSITS

The rivers of the world dump tons of material into the oceans each year. Sea floor springs and volcanic eruptions also add many tons of elements. Even the winds contribute solid materials to the oceans in appreciable quantities. Most of these sediments rapidly settle to the sea floor in near shore areas, in some cases forming potentially valuable placer mineral deposits. The dissolved load of the rivers however, mixes with seawater and is gradually dispersed over the total oceanic envelope of the earth. Because of the nature of the minerals and their mode of formation, it is convenient to consider the occurrence of ocean deposits in several environments—marine beaches, sea water, continental shelves, sub-sea floor consolidated rocks, and the marine sediments of the deep sea floor. Minerals are mined from all of these environments except for the deep sea floor, which was only recently recognized as a depository for mineral deposit of unbelievable extent and significant economic value.

CLASSIFICATION AND OCCURRENCE OF MINERAL DEPOSITS

Offshore mineral deposits can be classified into three categories :-

1. Dissolved minerals—sodium, magnesium, calcium, bromine etc
2. Unconsolidated minerals—Diamonds, gold, heavy metals etc
3. Consolidated minerals—iron ore, coal, sulfur etc

An estimated 10 tons of calcareous oozes, formed by the deposition of calcareous shelves and skeletons of planktonic organism, cover some 130 million square kilometers of the ocean floor. In a few instances, these oozes which occur within a few hundred kilometers of most nations bordering the sea, are almost due to calcium carbonate similar to that of the limestones used in the manufacture of Portland cement. Covering about 30 million square kilometers of the ocean floor in a great band across the northern and southern ends of the pacific ocean and across the southern ends of the Indian and Atlantic ocean are other oozes, consisting of the siliceous shelves and skeletons of plankton's animals and plants. Normally these oozes could serve in most of the applications for which diatomaceous earth is used, for fire and sound insulation, for lightweight concrete formulations, as filters, and as soil conditioner.

Of considerable interest are the sea floor phosphorite deposits on the coastal shelves of many nations. The phosphorite of California occurs as modules that vary in shape from flat slab a few metres across to small spherical forms termed oolites. The nodules are commonly found as a single layer at the surface of course grained sediments. Phosphorite composition from the California offshore area is surprisingly uniformed and contains potentially economically attractive amount of phosphorus. Another type of phosphate deposits has been discovered off the West Coast of Mexico. It contains as much as 40 per cent apatite, and some experts have speculated that upto 20 billion tons of recoverable phosphate rocks exists in the deposits.

An estimated 10 tons of red clay covers about 104 million square kilometers of the ocean floor. Although compositional analysis is not particularly exciting, red clay may possess some value as raw material in the clay product industry., or it may serve as a source of metals in the future. The average assay for alumina is above 15 per cent, but red clays from specific location have assayed as high as 25 per cent alumina; copper contents as high as 0.2 per cent also have been found. A few hundredths of a per cent of such metals as nickel and cobalt and a per cent or so of manganese also are generally present in micronodular fraction of the clays and in all likelihood can be separated and concentrated from the other materials by a screening process or by some other physical methods.

The remarkable deposits of Co-Ni-Cu-Mn nodules covering the deep ocean floors will require new concepts in materials handling and while some attempts may be made to mine them from the sea surface it is almost certain that future operations will include some form of manned equipment operating on the sea floor.

The discovery in 1978 of polymetalic sulphides at the mid-ocean spreading centres has aroused much interest, these sulfides include sediments enriched in iron and manganese. Sites of rich deposits have been located at the Galapagos spreading centres in the Gulf of California and at the East Pacific Rise. From an economic standpoint, the most interesting oceanic sediments are manganese nodules-small, black to brown, friable lumps found to be widely distributed throughout the major oceans.

Many theories have been proposed to account for the formation of manganese nodule, the best probably being that the ocean is saturated at its present state of acidity-alkalinity in iron and manganese. For this reason, these elements precipitate as colloidal particles that gradually increase in size and filter down to the sea floor. Colloids of manganese and iron oxides collect many metals and tend to agglomerate as nodules at the sea floor rather than settle as particles in the general sediments. An estimated 1.5 trillion tons of manganese nodules lie on the Pacific Ocean floor alone. Averaging about 4 cm in diameter and found in concentration as high as 38600 tons per square kilometer, these manganese nodules contain as much as 2.5 per cent copper, 2 per cent Nickel, .2 per cent Cobalt, and 35 per cent manganese.

Consolidated mineral deposits are contained within the bedrock underlying the sediments. These occur as in-place deposits and surficial deposits, the characteristics of which have a great influence on the economics of exploration and mining. Minerals such as coal, iron ore , limestone occur as surficial deposits. In-place deposits of coal, iron, tin, gold, and sulfur occur as massive, vein or tabular deposits.

EXPLOITATION

Once an ore body is indicated by geological geophysical, or other means, the next step is to sample it in area and in depth.

Mineral deposit sampling involves two stages. First, exploratory or qualitative sampling locates mineral values and allows preliminary judgement to be made. For marine deposits, this will involve such simple devices as snappers, drop corers, drag dredges, and divers. Second, the deposit must be characterized in sufficient detail to determine the production technology requirements and to estimate the profitability of its exploitation. This quantitative sampling requires much more sophisticated equipment than does the qualitative type, and for marine work few systems in existence can be considered reliable and accurate. However, in particular cases, systems can be put together using available hardware, which will satisfy the need to the accuracy required.

MINING

Dissolved Minerals

Commercial separation techniques for the recovery of minerals dissolved in seawater are limited to chemical precipitation and filtration for magnesium and bromine salts and solar evaporation for common salt and fresh water production on a limited scale. Other processes developed in the laboratory on pilot plant scale include electrolysis, electrodialysis, adsorption, ion exchange, chelation, oxidation, chlorination, and solvent extraction.

Unconsolidated Deposits

Unconsolidated deposits include all placer minerals, surficial and in place as well as the authigenic deposits found at moderate to great depths. The offshore mining of these deposits has become the most widely publicized facet of marine mining, largely because of the sudden awareness of the potential of manganese nodules as a mineral source and because of the important and exciting development in the exploitation of offshore diamonds in south west Africa.

So far the method of recovery which have been used or proposed have been conventional, namely by dredging using draglines, clamshells,bucket dredges, hydraulic dredges, or airlifts as illustrated in Fig 7.1. All these methods have been used in mining to maximum depth of 200 feet, and hydraulic dredges for digging upto 500 feet are being built. Dredging of such diverse products as diamonds, gold, heavy minerals sands, iron sands, tin sands, lime sands. The most important of all these commodities is the least exotic; 60 per cent of world production from marine unconsolidated deposit is involved in dredging and mining operations for sand and gravel. Other major contributors to world production are the operations for heavy mineral sand (ilmenite, rutile and zircon), mostly in Australia, and the tin operations in Thailand and Indonesia, which account for more than 10 per cent of worlds tin.

Consolidated deposits

The third and last area of offshore mineral resources has an equally long history. Undersea coal accounts for almost 30 per cent of the total coal production in Japan and just less than 10 per cent in Great Britain.

Extra costs associated with mining coal from under the sea have been due mainly to exploration, with mining and development being usually conventional. There is no doubt that costs will be greater, generally but on the other hand, in the initial years of offshore mining as a major industry, the prospects of finding accessible, high grade deposits will be greater than they are at present on land.

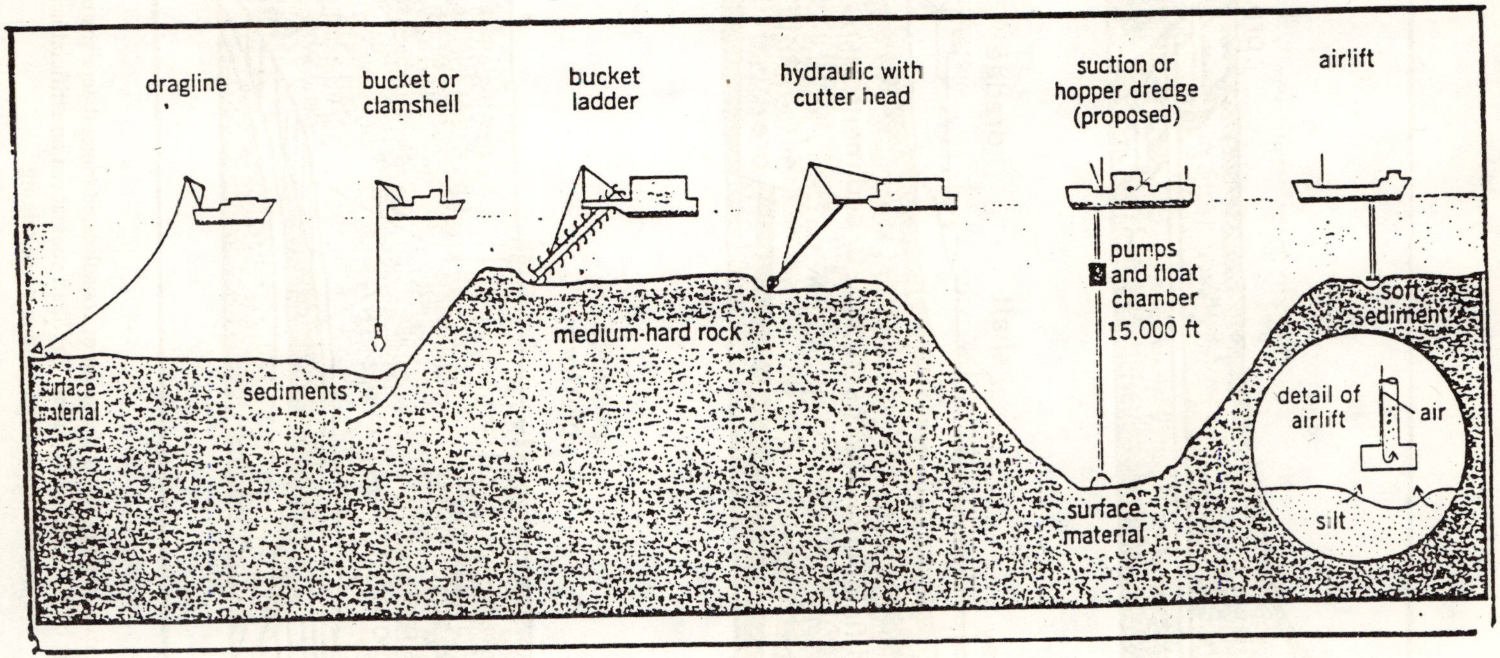

Fig. 7.1 : Methods of dredging used in the exploitation of unconsolidated mineral deposits offshore.

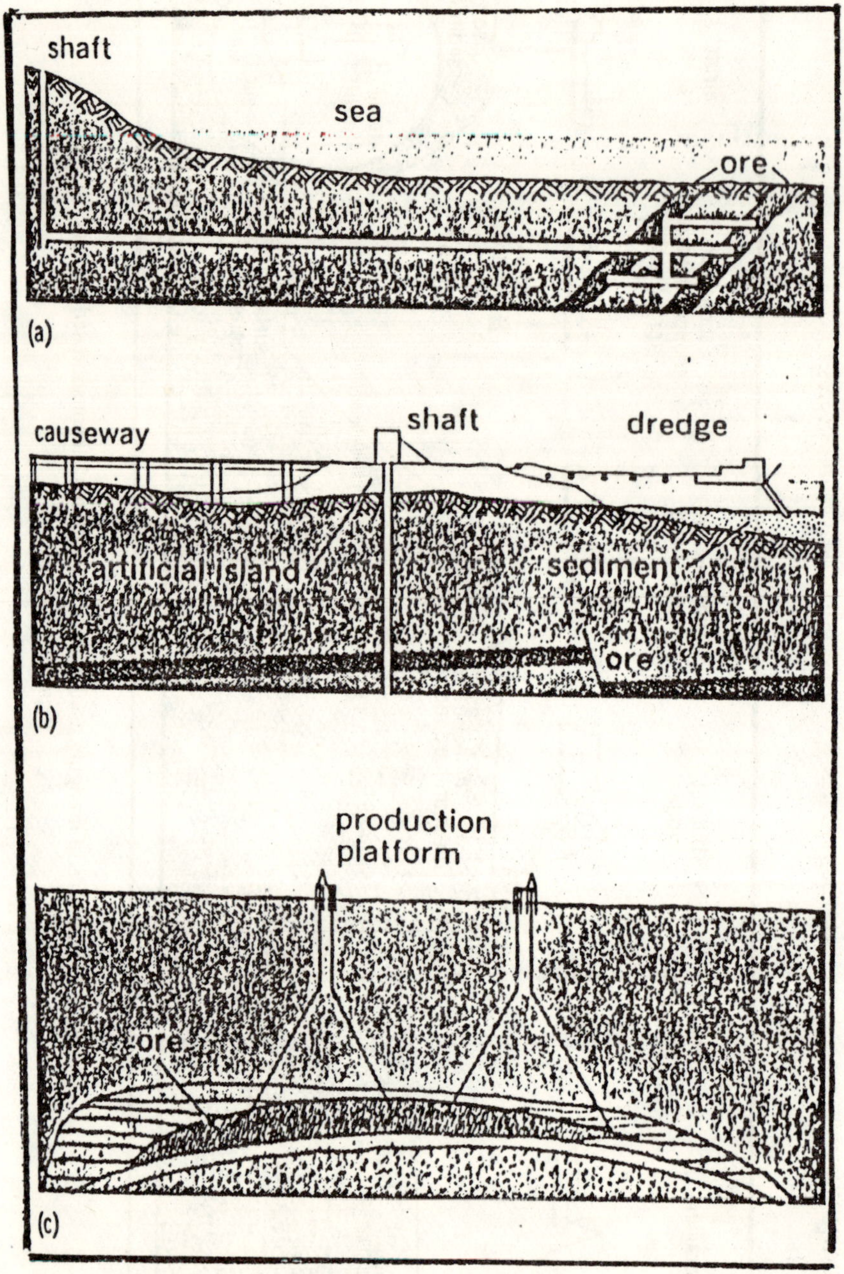

Fig. 7.2 : Methods of mining for exploitation of consolidated mineral deposits offshore (a) Shaft sunk on land, access by tunnel, (b) Shaft sunk at sea on artificial island, (c) Offshore drilling and in-place mining.

Some of the mining methods used are illustrated in Fig 7.2. For most of the bedded deposits which extend from shore workings a shaft is sunk on land with access under the sea by tunnel. Massive and vein deposits are also worked in this manner. Normal mining methods are used, but precautions must be taken with regard to overhead cover. Near land and in shallow water a shaft is sunk at sea on an artificial island. The islands are constructed by dredging from the sea bed or by transporting fill over causeways. Sinking through the island is accompanied by normal precautions for loose, waterlogged ground, and development and mining are thereafter conventional. Offshore drilling and in place mining are used only in the mining of sulfur, but this method has considerable possibilities for the mining of other minerals for which leaching is applicable. Sulpher, which is taken from salt domes in the Gulf of Mexico, is mined, by process in which pressurised hot water is pumped into the sulphur-containing cap of the dome, melting the sulphur and forcing it to the surface. Compressed air is also used to pump sulphur to the surface; the steel molten sulphur is then conveyed to the shore through insulated pipelines.

The Future

Despite the technical problems, which still have to be overcome, the future of the undersea mining industry is without doubt as potent as it is fascinating.

Deposits of hot metalliferous brines and oozes enriched with gold, silver, lead and copper have been located over a 38.5 million square kilometers area in the middle of the Red Sea at depths of 6000-7000 ft.Major problems of dissolved mineral extraction must be solved before their exploitation, and significant advances must be made in the handling of these corrosive media at such depths and distances from shore.

The mining of unconsolidated deposits will call for the development of bottom sited equipment to perform the massive earth moving operations that are carried out by conventional dredges today. The remarkable deposits of Co-Ni-Cu-Mn nodules covering the deep, ocean floors will require new concepts in materials handling, and while some attempts may be made to mine them from the sea surface, it is almost certain that future operations will include some form of manned equipment operating on the sea floor. Fig.7.3 illustrates the proposed system for mining deep sea nodule deposits using manned bottom vehicles.

Consolidated deposits may call for a variety of new mining methods, which will be dependent on the type, grade, and chemistry of the deposits, its distance from land, and the depth of water. The possibility of direct sea floor access at remote sites through shafts drilled in the sea floor is already being given consideration under the U.S.Navy's Rocksite programme and will be directly applicable to some under sea mining operations. In relatively shallow water, shaft could be sunk by rotary drilling with caissons. In deeper water, the drilling equipment could be placed on the sea floor and the shaft collared on completion.

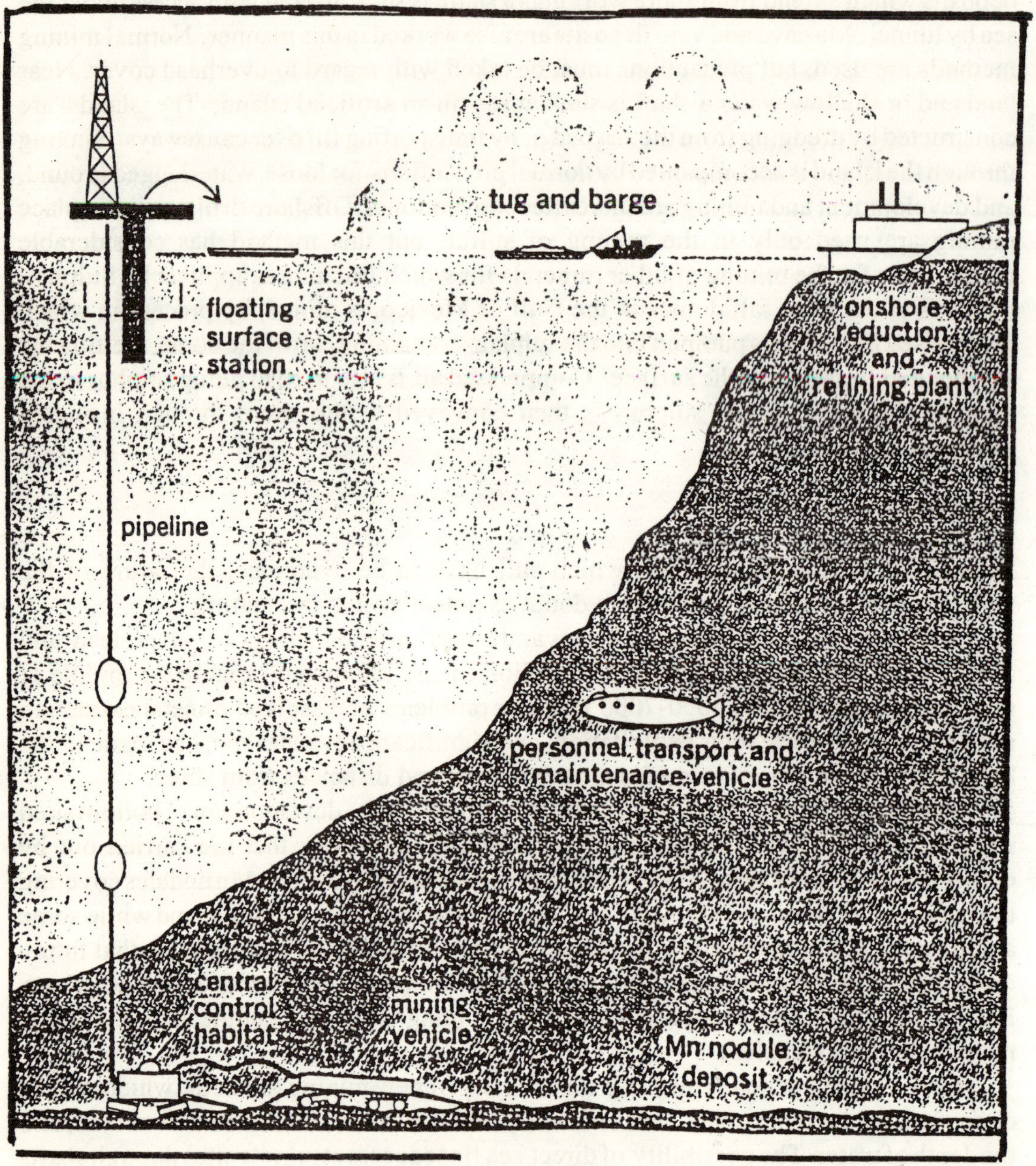

Fig. 7.3 : Proposed system for mining deep-sea nodule deposits using manned bottom vehicles. A nodule density of 2 lb/ft^2 will require daily coverage of 115 acres for a production of 5000 tons/day. (U.S. Bureau of Mines)

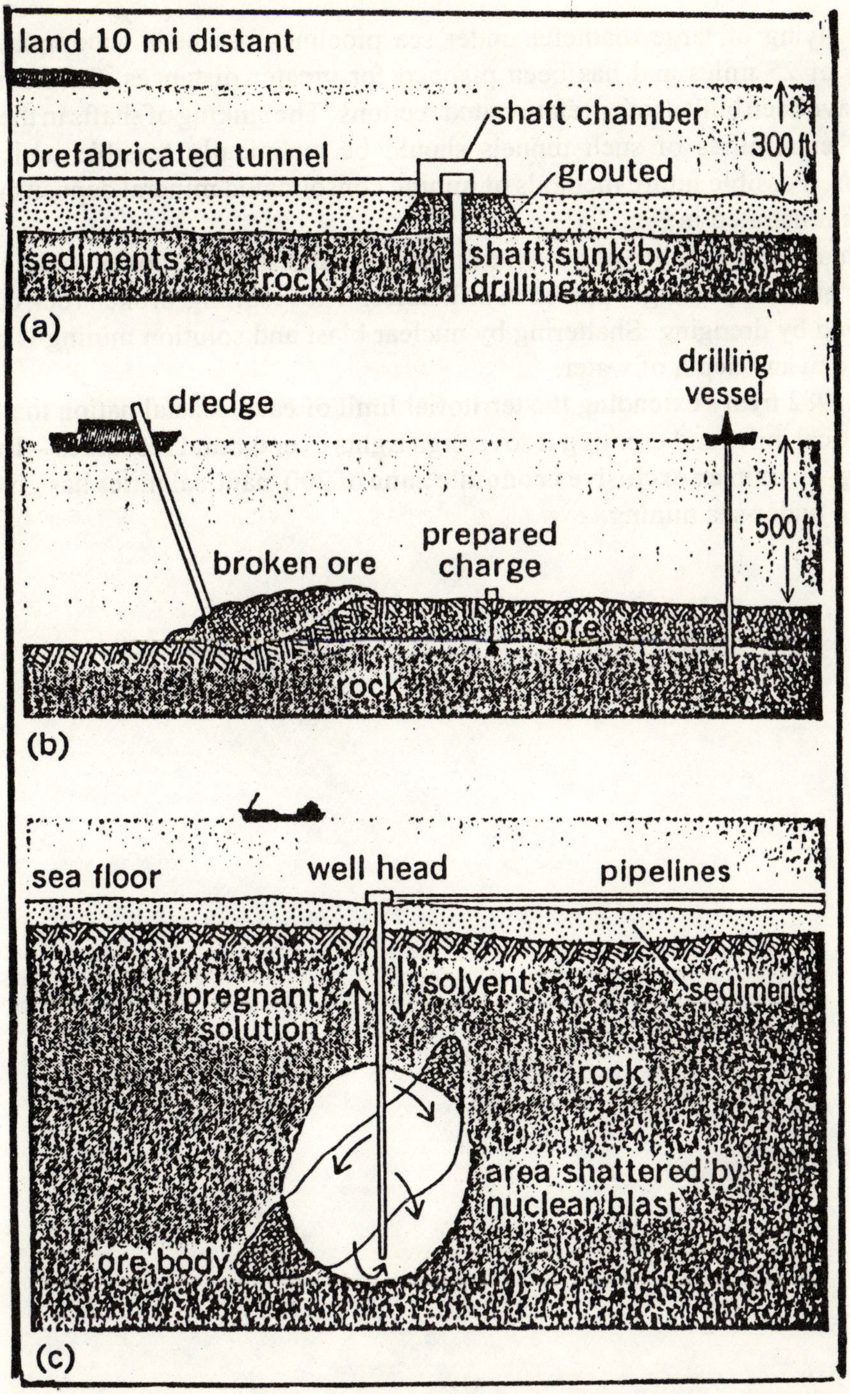

Fig. 7.4 : Possible future methods of mining consolidated mineral deposits under the sea, (a) Shaft sinking by rotary drilling from tunnels laid on the sea floor, (b) Breaking by nuclear blasting and dredging, (c) Shattering by nuclear blast and solution mining.

The laying of large diameter under sea pipelines has been accomplished over distances of 25 miles and has been planned for greater distances. Substuarine road tunnels have been built using prefabricated sections. The sinking of shafts in the sea floor from the extremities of such tunnels should be technically feasible under certain conditions. Possible future methods of mining consolidated mineral deposits under the sea are as illustrated in Fig. 7.4.

Submarine ore bodies of massive dimensions and shallow cover could be broken by means of nuclear charges placed in drill holes. The resulting broken rock could then be removed by dredging. Shattering by nuclear blast and solution mining is a method applicable in any depth of water.

The 1982 treaty extending the territorial limit of each coastal nation to a distance of 12 nautical miles and granting it sovereign rights over natural resources—**living and nonliving**—within an **exclusive economic zone** of 200 nautical miles has given a new impetus to under-sea mining.

III

PORTS AND OTHERS LOGISTICS

EXPLOITATION OF OCEAN RESOURCES—SUGGESTED MODALITIES, FOR ELECTRIC POWER GENERATION, FROM SEA WAVES

8

Sri Ch. Sree Rama Sarma,
M. E., F. T. E. (India).,
Ex-Executive Director,
A. P. S. E. Board, Hyderabad.
Presently, Member, C.B.I.P., Govt. of India.

ABSTRACT

Information about offshore sea waves, forms interesting parameters in transfer of energy, propagating in different directions. Recent studies, have shown distribution of wave energy as a function of frequency, for which there is need to establish Indian Coastal Waters characterised by different depths and foreshore slopes. India's coastal length is in the order of about 600 km, offering an estimated power potential of about 60,000 MW of electrical Energy. Using the power of Sea wave has been tested since ages through Ram Air Generator, Oscillating Water Column and similar means. The demonstration model now presented in this paper, has *a float converting wave energy into stored energy through a driven gear wheel unit. This energy is used to charge a double coil wound spring drum. A small experimental set-up of a electric power generator model is discussed in this paper, with the coil spring as its prime mover.* This Sea Wave Generator is highly Reliable and operates round the clock, with good efficiency and economy, and deserves all encouragement. A short Audio-Video presentation available, with the author will be projected during presentation of this paper.

INTRODUCTION

The sea wave have perennial energy. As waves touch the shores abundant energy is released. Wave energy conversion to electrical Energy is the most efficient source of energy. According to experts the ocean offers power potential of 60,000 MW along India's 6000 km coast line. A simple energy transfer scheme is suggested as an economical way to produce electric power for use in isolated seashore colonies and far off towns. Electricity may be provided for village water supply, well water pumping for agriculture, live stock water supply, refrigeration plants for medical clinics, crop drying, preservation, grass cutting and other similar units through the Sea Wave Generator Plant. It is a more Energy-Efficient Technology than all the prevailing non-conventional sources and increases the hope for optimum and fullest use of Renewable Energy Resources.

PLANNING

Wave energy systems are classified into three main categories, namely:

(i) Coastal
(ii) Offshore floating, and
(iii) Offshore submerged.

Every square meter of ocean can give possible energy content of 10kw/m front. Forecasting energy potential along the coastline, location of plant site, developing reliable design for the plant depending upon the location and the pricing policy and administering the new energy technology programme for meeting the energy shortage, will all form the salient features for successful implementation of Sea Wave Power Generator Scheme. Considering the vast Indian coastline and the potential wave energy available, which is yet to be explored, it is necessary to develop a low cost model of a sea wave power generator. There is ample scope for installing 5 kw - 10 kw units in many of the coastline villages. The proposed unit shall be made with indigenous materials only. For the 21st Century, the sea wave generator is the most efficient and economical non-conventional source of electrical energy, with Several advantages, as follows :

- Low cost power.
- Availability of wave energy round the clock in all seasons, unlike solar and wind energies.
- Pollution-free, Non-hazardous.
- No fuel or any raw material inputs.
- Capital costs and running are the lowest amongst all other alternate sources of power, viz. Solar, Wind, Biogas etc, and most importantly.
- Practically negligible maintenance.

SPECIFICATIONS

A one metre amplitude wave of 10 seconds period will give a power output of 40 kw per meter of wave front. Analysis of Ocean Weather Ship India data shows that power levels of 30-40 kw per metre are most frequent, with some power levels of 80 kw per metre. The sea surface in some respects is like a huge wind mill. Ocean waves transfer energy with high efficiency, though the attenuation distances are very large. Fourier analysed a set of tidal waves with different wave lengths, phase and directions. The power (P), associated with a small amplitude sinusoidal wave of single wave length, also called monochromatic wave, is given by :

P = ((E * V) / 1 metre of wave front) kw. where,
E = Total energy in the wave per unit surface area
V = Phase Velocity.

The design of the device and the energy that can be extracted from the sea wave depend upon the directionality of the wave due to its widely spreading characteristics while approaching a point. The Indian coastal waters are characterised by different water depths, fore-shore slopes and bathymetry. The spectral distribution of wave energy can provide valuable information about the range of frequencies over which significant amount of wave energy, spread and the predominant frequencies in the wave train where maximum energy is concentrated. Output from the wave is highly variable over a number of periodic cycles. It is possible to average the output of Sea Wave Generator. If "Et" represents the total energy of the wave the average R. M. S. value will be (Et/2). However, since the float operates only for a period of two quarters of a full cycle wave, advantage is derived in retrieving total energy in the forward rising motion and the downward retrieving motion of the wave by using full wave float

UNIT FLOAT

In order to design the most suitable gear transmission unit, it is essential that a through knowledge of the characteristics of the driven machine, are known. The approximate service factor must be determined by considering the total operational hours per day, the number of starts and stops per hour and the load classification, which may basically be determined by calculating the Mass Acceleration Factor (MAF). For an uniform load, the MAF may be around 0.2, whereas for moderate shock loads, the permissible MAF may go up to 10.

In the classification of load for the design evolved, the data adopted is reviewed. A duty of 200 starts and stops per hour, running at 24 hours per day, is taken as loading cycle.

A direct-drive power generator gives its maximum power output at a very low speed, say 180 RPM. For units with outputs of a bore 5 kw, it is much larger, heavier and more expensive than the equivalant gear driven units. The advantage of using a direct-drive generator is that the lower speeds mean longer life and less wear and tear of the generator bearings.

DEVELOPMENT OF MODEL

For the wave energy to be usefully employed the device must accept the wide range of frequencies and amplitudes presented by the sea. A float system is modelled to respond to surface motion of the water. In doing so, the up and down reciprocal motion of sea waves is converted directly into continuous rotary motion. *The output is realised as stores energy through a continuously charging spring. The stored energy will be used to drive a low speed generator.*

The operating sequence is shown in the Block Diagram at Annex 1.

A 20 w model was successfully tested on the sea cost of A. P. State. This design is already patented in the U. S. and in India by an enterpreneur.

There is scope for more extensive utilisation of the device on a commercial scale.

SALIENT FEATURES

The up and down reciprocal motion of sea waves is converted directly into rotary motion by means of a buoyant cylindrical drum. The drum will be floating to the lift given by wave energy and the output is transferred to continuously charging strap spring for acquiring stored energy, which is consequentially used to drive a prime mover of an Electric Power Generator. A governor speed control is coupled to drum through a clutch. A power transmission device comprises spring mechanism fixed to sprocket gears.

This simple unit proves to be an economical way to generate electricity in isolated seashore colonies. In future, the Seawave Generator is bound to be the most economical aMumbaind efficient non-conventional source of electrical power.

The Advantages are :

(i) Cheapest power source, due to long generating hours without any fuel, whatsoever.
(ii) No requirement for pollution controls.
(iii) No problems, like evacuation of products of combustion like ash etc.
(iv) No raw material requirements.
(v) Free Source of Seawave power.
(vi) Capital cost and running costs are the lowest, even amongst other alternate sources of power i.e., Solar and Wind etc.

The Sea Wave 'Power Transmission Device', is registered under U. S. patent No. 4,832,155. The device is also registered under Govt. of India, Patent Act, with Registered No. 163607.

COST ECONOMICS

For an experimental study, the cost of a 10 kw generator, power transmission device, mounting scheme, structures and buoyant cylinder, works out to Rs. 3.50 lakhs. Considering a selling price of Rs. 2.50 per unit the pay back period for the capital cost work out to 2.00 years. A 1 kw low speed generator will be receiving the drive from the proposed mechanical driving systems.

Details are furnished below :

Cost

10 kw. Seawave Generator Plant.

S. No.	*Particulars*		*Amount Rs. Lakhs*
1.	10 kw A. C. Generator		0.30
2.	Spring mechanism, governor control and power transumission device		0.40
3.	Foundation, Frame and Structures		1.50
4.	Electrical Switchgear 4 lines 1 kw		0.40
5.	Miscellaneous Expenditure		
	Consultancy	3%	
	O & M	1%	
	T & P	1%	
	Transport and Labour etc.	27%	0.83
	Total		3.43
	Say, Rs. 3.5 Lakhs.		
	Cost/kw = Rs. 35,000		

Revenue Return

Output per annum :

10 kw × 22 hrs. × 350 days = 77,000 kw/hrs.
Re. 2.50 × 77,000 = Rs. 1,92,500
Return on Capital Cost = 3,50,000/1,92,500 = 2 Year (Approx.)
Pay Back Period = 2 Years.

A comparison of economical aspects with reference to other sources of Non-conventional and conventional sources of energy, on a large scale, is given in Table 8.1.

CONCLUSION

This plant can be tailor-made to suit the site conditions at various locations of the coastline and put to immediate beneficial use. The cost per kw output, with a number of installations on one platform in cascade operation, works out to Rs. 35,000/-, or Rs 3.50

crores per MW. Considering that this price compares favourably, even with that of conventional and other non-conventional sources of energy, e.g., solar, wind, nuclear, gas, mini-hydel etc., the Sea Wave Power Generator is worth implementation on a large scale and deserves all encouragement, by one and all.

REFERENCES

1. I. E. E. Publications.
2. C. B I. P. Journals.
3. S. E. W. Manual.
4. Naduri. V. Dr. Sea Wave Power Generation Technical Report.
5. Raju K. V. S. T. Technical Report.

ANNEX 1

BLOCK DIAGRAM - OPERATING SEQUENCE

Fleat/Raft

Drive Wheel - 1

Drive Wheel - 2

Charged Spring

Assembly ↓

Clutch Control

Rachet Gear

Buffer Spring

Storage

Sprocket

Gear Wheel - 3 ↓

Governor Control

A. C. Generator

Battery Bank

Charging Unit

Switch Control

Gear at Mains. ↓

TABLE 8.1 : **Comparision of Economical Aspects**

Source	*Estimated Running Hours Per Annum.*	*Cost Per KW (Plant+Lines)*	*Estimated Cost/KW Including Transmission and Distribution Lines*	*Remarks*
Sea Wave Generator.	8560	Rs. 35,000 + 2,000	Rs. 37,000/-	(i) Installation and running costs are cheapest. (ii) Availability throughout the year. (iii) Constant power out put.
Solar	2160	Rs. 35,000 + 2,000	Rs. 37,000/-	(i) Working hours limited due to Night, Monsoon days etc.
Wind	1600	Rs. 30,000 + 2000	Rs. 32,000/-	(ii) Maxm. Availability 8 hrs/day. Constant output not possible due to varied wind velocities.
Thermal	8560	Rs. 20,000 + 15,000	Rs. 35,000/-	(i) Lot of pollution. (ii) Output depends on coal availability. (iii) Disposal of Ash problematic.
Hydel	5840	Rs. 20,000 + 15,000	Rs. 35,000/-	(i) Long gestation period. (ii) Constant output not possible due to vagaries of monsoon.
Nuclear	8760	Rs. 25,000 + 15,000	Rs. 40,000/-	Lot of Pollution and nuclear hazards.
Gas	8560	Rs. 15,000 + 20,000	Rs. 35,000/-	(i) Lot of pollution problems. (ii) Output depends on gas availability.

NON-LIVING OCEANIC WEALTH OF INDIA

9

Prof. G. S. Roonwal
Department of Geology University of Delhi, India

INDIA AND THE SEA AROUND US

In the coming decade we will look increasingly to the oceans for many of our non-renewable mineral resources. One would understand that the development pace may go up or down depending upon the mineral prices in the world market as well as the discovery of new onshore mineral resources. We know that oil and natural gas worth billions of dollars per annum are extracted from the offshore zone. In India, offshore petroleum recovery accounts for more than half of the indigenous petroleum products.

But we must remember also that these riches of the seafloor do not come readily. Exploration of the sea shore and offshore deposits require high technology and big capital as well as extensive planning. One may classify oceanic resources into three types.

OFFSHORE BED ROCK DEPOSIT

These deposits are essentially seaward equivalent of onshore deposits, which include hydrocarbons, coal, metalliferous deposits, sulphur and evaporites. They could extend offshore for hundreds of kilometres and down to depths of thousands of metres below the sediment water interface. For the most part deposits of this type are exploited by modification of onshore type operation such as the use of various types of drilling platforms in the case of, say petroleum, such as in the Bombay High Zone. Drilling and solution mining are used for sulphur (as in Mexico) and could be used for offshore extraction of evaporites, especially common salts and potash. Subsea mining of coal from shore developed operations has been underway for many years in Japan, the U. K. and Canada. Similarly, metalliferous deposits have been mined offshore in many parts of the world. Also we should remember that much of the world oil and gas production which is located offshore, occurs in the Exclusive Economic Zone (EEZ) and contributes a significant national resource

DEEP-SEA OCEANIC MINERAL DEPOSITS

Deep oceanic mineral deposits are found in the open sea down to abyssal depths usually far from the land where chemical and biochemical sediments are not diluted by input of land derived terrigenous material. No deep oceanic deposits have been commercially exploited to date. There have been various attempts to commercially evaluate them. It is evident from these that the development and mining cost will be enormous. Up to now the main interest has been focused on manganese nodules (also called polymetallic nodules) which are found in many parts of the deep ocean at abyssal depths. Presently, interest has grown in the volcanogenic massive sulphides (VMS) found as clusters of chimneys at several locations in the mid-oceanic ridges as well as the cobalt-rich crusts found on many of the ancient volcanic seamounts in the area of upwelling. The commercial interest in manganese nodules is not in manganese content, but rather in the nickel, copper and cobalt contents. In the areas of the northern Pacific Ocean the copper + nickel + cobalt content in nodules is about 3 per cent and manganese around 25 per cent to 30 per cent. It is in this Pacific Ocean area that much of the commercial interest has been focused by countries like USA, Germany, Japan, China, Russia, France and Korea. Research, exploration and tests have been on mining technology and environmental impacts assessment have been conducted by the USA, Germany and Japan. Korea and Chian are making special efforts to conduct such a study. In the Indian Ocean, manganese nodules occur in several locations which are defined in each oceanographic basin. However, the best grade nodules occur within the Central Indian Basin. It is in this basinal area that Indian efforts have been made. As a result of this, the UN International Seabed Authority (ISA) has granted India "Pioneer Investor" status. Indian efforts are aimed towards mining these nodules.

Volcanogenic massive sulphides (VMS) in the Indian Ocean were first located in December 1993 near north of the 'Triple Junction' area of the mid-Indian Ocean Ridge. Evidences of such sulphides occurrences were though known. This again states that oceanic research is not an easy task. These deposits are located in the "fourth Central Indian Ridge segment", close to the "rift axis". This occurrence is now not an active field. There is a strong likelihood of VMS deposits occurring in the Andaman Sea (EEZ of India) where also evidence of VMS have been obtained.

Metalliferous muds and brines of the Red Sea are also a type of sulphide resource but with somewhat different origin in that the hot metal brines are derived from the circulation of ground waters through evaporation of the Red Sea graben resulting in sediment and brine rich metals such as copper, zinc and silver.

A relatively new interest appears in the ferromanganese encrustation/crusts because of higher levels of cobalt contents in them. In addition, they also have high levels of platinum and ruthenium. These crusts best occur on many of the ancient seamounts and are found in the open sea, mainly on the upper slopes and marginal plateaus, at depths of about 1000 - 2000 mts. rather than at depth of 4000 - 6000 metres where nodules characteristically occur. In fact this relatively shallower water depth may make them commercially more attractive but unlike manganese nodules, they are likely to be irregular in their occurrence and difficult to extract from the seamounts. However, many of the locations would fall well within the EEZ of India.

In summary, there is no doubt that manganese nodules, sediments, and brines of oceanic depth constitute an important potential resource for the future. But there is little possibility of them being exploited immediately because of a combination of factors such as low metal prices, recent success in landbased deposits, and widespread substitutes of many metals by cheaper materials. Indeed, it may be that some of the oceanic oozes, such as silicon ooze, may ultimately have more commercial relevance because of their physical properties, than the more glamorous metalliferous nodules of the oceanic depths.

SHALLOW OCEANIC MINERAL DEPOSITS

It is towards the sea shore that greatest efforts in India would probably be directed where with sea level changes, environment, tectonic, and resource evolution is achievable. In addition to the oil and gas, a number of marine commodities occur in the coastal zone—a resource in its own right which is significant, because of mineral sands (placers) of titanium, rutile, zircon, ilmenite, magnetite and monazite in them. Construction materials such as gravel, sand, carbonates and phosphorites used in agriculture, are being extracted from the sea floor at the present time.

THE DESIGN OF WELL PLATFORMS IN WESTERN OFFSHORE

10

P. K. Johri
Oil and Natural Gas Commission

ABSTRACT

The Western Offshore of India is a major source of hydrocarbons for the country with its offshore oil and gas fields such as Bombay High, Neelam, Heera, Bassein and other marginal fields. The area has been under focus for intensive field development, capital expenditure and technology upgradation. In the past two decades over 150 well and process platforms alongwith vast pipeline network were installed for exploitation of hydrocarbons from the different fields in Western Offshore.

A crucial aspect of development activities has been the engineering of these facilities, which is multi-disciplinary in nature and is generally completed in two stages: Basic and Detail. Steel template structures comprising of a jacket and two-level deck (topside) have been used for all the well-platforms keeping in view the operational and environmental criteria and process parameters which influence the deck layout, process facilities and utilities. These platforms facilitate drilling of wells, well-completion, well-testing, and transportation of well-fluid to process platforms.

The paper discusses the design aspect of well-platforms and brings out the salient features at various stages. Further, the changes that have taken place over the years as a result of optimisation of topside facilities, advancement of technology etc., and their impact on deck size and cost are outlined.

INTRODUCTION

The Western offshore of India is a major source of hydrocarbon for the country with its offshore oil and gas fields such as Bombay High, Neelam, Heera, Bassein and other marginal fields. The area has been under focus for intensive field development, capital expenditure and technology upgradation. In the past two decades over 150 well and process platforms along with vast pipeline network were installed for exploitation of hydrocarbons from the different fields in Western Offshore.

A crucial aspect of development—activities has been the engineering of these facilities, which is multidisciplinary in nature and is generally completed in two stages: Basic and Detail. The salient features of design of well platforms installed to facilitate drilling of wells, well completion, well testing and transportation of well fluid to process platforms, during both basic and detail engineering stages are brought out in the following paragraphs:

Basic Engineering

Figure 10.1 depicts the stages involved in the overall scheme of project execution. The important input for development of conceptual report is the technological scheme with detailed development plan and consisting of geological inputs covering reservoir studies, exploratory drilling parameters, process parameters, number of development-wells and their location, approximate location of well platforms and process platform, gas lift and water injection requirement, evacuation plan of oil and gas etc. Basic engineering enables preparation of bid documents which cover designing/sizing of platform facilities and pipelines, costs and economics of development. Basic engineering enables the bidders to understand the scope of work, detailed specifications, drawing and schedule etc. and facilitates submission of bids that are complete in all respects.

The Technological Scheme contains the following details:

- Number of wells (initial, final)
- Number of single or duel completion (initial, final)
- Number of injectors (initial, final)
- Year-wise and well-wise production, artificial lift requirement
- GOR, water cut and pressure
- Optimum water injection pressure at each well head
- Optimum gas injection pressure at each well head
- Number of well-platforms and process platforms
- Water depth, reservoir depths
- Composition of gas/well-fluid/produced water
- Quality of injection water
- Completion details of wells.

These details are firmed up based on exploratory drilling and reservoir studies and are used in preparing the Conceptual Study Report (CSR). The CSR presents an engineering scheme to achieve the objective of field development and covers preliminary sizing of facilities, cost and economics of the scheme development. Some of the important issues covered in the CSR are:

1. The well-wise and year-wise production, GOR, Water cut and pressure of all wells earmarked for drilling through well platform are combined to get the year-wise production of a platform.
2. The year-wise and well-wise gas lift requirement of each well is then added to work out total oil and gas handling for each platform.
3. Similarly, the year-wise injection water requirements are also worked out for each platform.

DESIGN OF WELL PLATFORMS IN WESTERN OFFSHORE

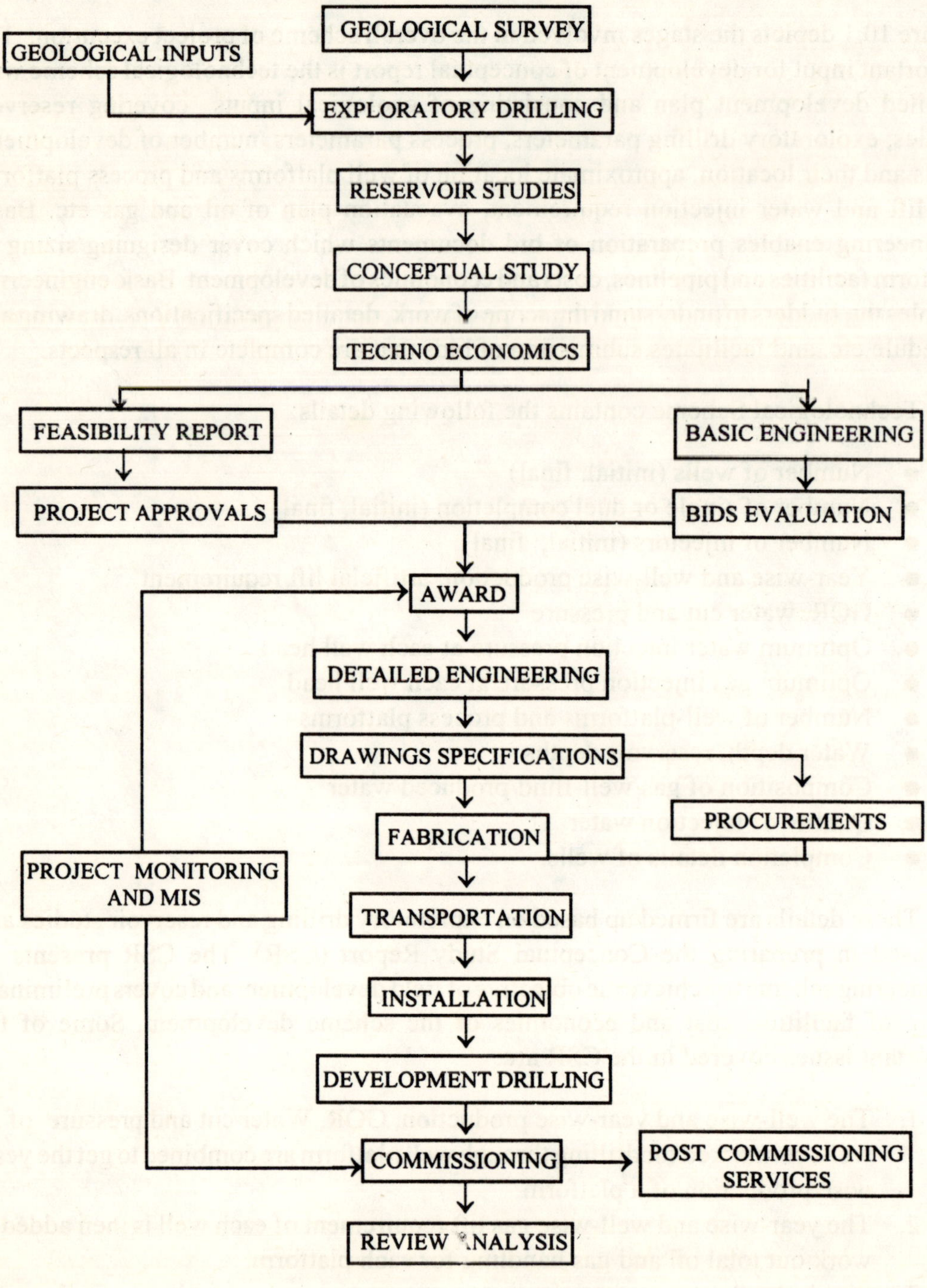

Fig. 10.1: Scheme of Project Execution

4. The platform-wise tubing head pressures and temperatures are also worked out based on reservoir pressure and well completion details.
5. Size and approximate route of oil, gas lift and water injection pipelines between central process platform and well head platform.
6. Details (including metallurgy for handling oil/gas) and size the facilities on well-head platform based on the company policies, processing requirements and statutory guidelines.

The CSR thus form the basis for preparation of Feasibility Report and Basic Engineering. Feasibility report addresses the issue of economic viability of the proposed development plan and is prepared for submission to statutory agencies for approval. Basic engineering results in the preparation of bid package for the project.

Bid package for a specific project comprises of two parts:

(i) Commercial part dealing with all aspects related to execution of contract including payment, instructions to bidders, bid evaluation criteria and general conditions of contract,
(ii) Technical part which essentially clarifies the scope of work involved in the project.

The technical part includes design criteria for various disciplines, technical specifications for different equipment/items, requirements with respect to procurement, fabrication, transportation, installation, pre-commissioning and commissioning, drawings such as P&IDs, PFDs which are the results of process engineering carried out based on inputs from CSR, Single Line Diagram (SLD), pipeline route drawings and proposed structural design covering the different loads that need to be taken into consideration. Extensive use of applicable codes (Recommended Practice/Standards) of API and other industry standards is made to specify the design/safety requirements.

Basic engineering, thus accomplishes the twin tasks of:

(i) Documenting the scope of work involved in the project, discipline-wise,
(ii) Enabling submission of offers from bidders complete in all aspects to facilitate better bid evaluation, minimise conflict resolution, and reduce deviations from estimated costs.

DETAIL ENGINEERING

This is the concluding phase of engineering and the objective is to finalize/freeze all the process parameters, structural design/configuration, specifications for various equipment, sizing calculations, drawings and other related outstanding issues to facilitate commencement of fabrication and placement of orders for required equipment. A typical list of major equipment installed on a well platform is placed at Annexure-1. The salient activities in this phase discipline-wise are as given below:

(i) *Structural:* Steel template structures comprising of substructure including jackets/ piles and topside consisting of cellar deck, main deck, heli-deck are generally designed. The major issues finalized are:

- Loads for which the structure is to be designed:
  (Dead loads, Live loads, environmental loads, pre-service loads, accidental loads).
- Design of jacket structure including:
  Jacket in-place analysis, selection of jacket and pile configuration, framing fatigue analysis and foundation design.

The design is ensured to be in line with the provision/practices laid down in API & DNV codes.

Finalisation of:

Process	P&IDs, PFDs Equipment Sizing, Simulation of process
Piping	General arrangement Drawing (GAD) Isometrics Specifications for piping equipment
Mechanical	Equipment Layout Specification for Rotary/Safety Equipment
Instrumentation	Fire and Gas Detection System Control Philosophy Emergency Shutdown system Specification for Equipment
Electrical	Area Classification Diagrams Single Line Diagrams Cathodic Protection System Specification for equipment
Pipeline	Pipeline sizing Route diagrams Specification for material procurement.

In addition, basis for monitoring progress, reporting and planning for timely completion of the project are finalized.

RECENT DEVELOPMENTS

Since the mid 1970s when the first well platforms were installed, the jackets and decks have shown a steady trend of increase in size and cost necessitating a review/study of the rationale behind the provision of certain facilities. Some of the important reasons for this increase were:

- Increase in number of wells resulting in increase in inlet manifold size
- Need for secondary recovery facilities like water injection and gas lift
- Additional facilities for sour production

The impact on the cost/platform can be gauged from the fact that the deck size (including cellar and main decks) increased from 325 sq.m in 1976 to about 1000 sq.m in 1988 and the corresponding weight from 300 MT to over 850 MT.

The need to simplify/optimize topside facilities was recognised and efforts were made to derive the benefits of:

- Reduction in cost
- Reduction in execution time
- Indigenisation

These efforts resulted in resizing of process facilities and utilities and also in deletion of certain non essential equipment. The resulting deck facilities after resizing and deletion were ensured to be conforming to the provisions of various applicable codes and standards prior to implementation of the changes. Details of implementation carried out in phases are placed at Annexure 2.

Thus, while the underlying methodology of design has been relatively unchanged, recent developments in design focussed on simplification/optimization of topside facilities to reduce the overall cost/platform.

CONCLUSIONS

An important outcome of the involvement of E&C Division in the development of offshore oil fields in the Western Offshore of India has been the acquisition of indigenous engineering capabilities. This is evident from the fact that engineering for most well platforms / pipeline projects has been successfully carried out by the in-house Engineering (E&P) Section. Many more such projects and others involving innovative techniques such as installation of clamp-on structures for slot-addition on existing well platforms are in the pipeline.

Given the thrust on deep water development and exploitation of marginal fields, the future presents both challenges and opportunities in this area. In light of the available/existing capabilities and the track record of the section in not only assimilating new techniques but also in adapting them to meet field-specific requirements, the prospects of the Section in evolving into an **independent engineering consultant** for development projects in offshore oil industry, both in India and abroad, appear bright.

ANNEXURE 1

LIST OF MAJOR EQUIPMENT (TYPICAL WELL PLATFORMS)

- INLET MANIFOLD
- WATER INJECTION MANIFOLD
- GAS LIFT MANIFOLD
- TEST SEPARATOR
- OCI/GCI TANK
- OCI/GCI PUMP
- LAUNCHER/RECEIVER
- DECK CRANE
- DIESEL STORAGE TANK
- IUG SYSTEM
- IG RERCEIVER
- VOLUME BOTTLE
- G/S POT
- DCP SKID
- SD PANEL
- VENT BOOM
- CC PUMP
- CC DRUM
- OCI/GCI DRUM RACK
- BATTERY ROOM
- TELEMETRY ROOM
- SWITCH GEAR ROOM
- EMERGENCY CAPSULE
- SOLAR PANEL

ANNEXURE 2

DELETION OF ITEMS—PHASE I

CHEMICAL STORAGE TANK AND CHEMICAL INJECTION PUMPS
U/V STERILIZERS
POTABLE WATER TANK
SALT WATER TANK
BUNK HOUSE AND A/C
OXYGEN SCAVENGERS AND BACTERICIDE INJECTION PUMP
MONORAIL

REDUCTION OF CAPACITY DUE TO OPTIMISATION OF PLATFORM IN PHASE I

1. CRUDE CONDENSATE DRUM	SIZES FOR 7 DAYS INSTEAD OF 15 DAYS
2. INSTT. GAS DRIER	REDUCTION IN SIZE
3. TEST SEPARATOR	SIZED FOR 2500 BLPD INSTEAD OF 5000 BLPD
4. DIESEL GENERATOR SET	REDUCED TO 37 kW FROM 55 kW
5. EMERGENCY BATTERY	REDUCED TO 3000 AHR FROM 4500 AHR
6. L.T. SWITCH GEAR	REDUCED TO 100 A FROM 200 A

PHASE II

1. WORK STARTED	DECEMBER 1989
2. REPORT SUBMITTED	MARCH 1990
EQUIPMENT DELETED	5 NOS.
REDUCTION IN SIZE / CAPACITY OF EQPT.	13 NOS
3. REDUCTION 1 PLATFORM AREA	250 M2 APPROX.
4. REDUCTION IN STRUCTURE STEEL	100 MT APPROX.

COST SAVING: APPROX. RS. 3.15 CRS. PER PLATFORM

DELETION OF ITEMS IN PHASE II

1. SUMP CAISSON
2. I/U AIR COMPRESSOR IN PLATFORMS WHICH ARE PRESENTLY SWEET.
3. SPACE FOR FUTURE LAUNCHER/RECEIVER
4. NITROGEN BACK UP FOR INSTRUMENT AIR AND FIRE WATER PUMP START UP.
5. FQS IN GAS LIFT INSTRUMENTATION (FLOW TOTALIZERS)

REDUCTION IN CAPACITY / SIZE OF ITEMS DUE TO OPTIMISATION OF PLATFORM IN PHASE II.

1. CRUDE CONDENSATE DRUM	REDUCED TO 2 M3 FROM 10 M3
2. DIESEL STORAGE TANK	CAPACITY REDUCED TO 2 M3 FROM 10 M3
3. CRUDE CONDENSATE PUMP	CAPACITY REDUCED TO 300 LT/HR.
4. INSTRUMENTATION	NO. OF RTU REDUCED TO HALF
5. FIRE WATER PUMP	CAPACITY REDUCED TO 325 M3/HR FROM 450 M3/HR.
6. GAS DETECTORS (HC & H2S)	NO. REDUCED TO 13 & 12 FROM 25 EACH (HC & H_2S)
7. DG SET	CAP. REDUCED TO 10 kW FROM 37 kW
8. EMERGENCY BATTERY AND BATTERY CHARGER	CAP. REDUCED TO 1500 AH FROM 3000 AH
9. GAS DETECTION BATTERY, BATTERY CHARGER AND SOLAR PANEL	60% REDUCED FOR SWEET FIELDS, 30% REDUCED FOR SOUR FIELDS
10. PLATFORM LIGHTING	LOAD REDUCTION IN LIGHTING TO 7-8 kW FROM 17-18 kW.
11. L.T. SWITCH GEAR	RATING REDUCED TO 25 A FROM 100 A.
12. FIRE WALL IN WH AREA	REPLACED BY STEEL ISOLATING WALL
13. TEST SEPARATOR	CAP. CHANGED TO 3000 BLPD.

PHASE III

1. STUDY FOR MOC AND CORROSION ALLOWANCE FOR PLATFORM AND PIPELINE	JANUARY 1990
2. REPORT SUBMITTED	MAY 1990
3. COST SAVING	APPROX. RS. 2.8 CRS. PER PLATFORM

OPTIMIZATION IN PHASE I, PHASE II AND PHASE III HAVE BEEN IMPLIMENTED IN ABOUT 37 WELL PLATFORMS WHICH HAS ALREADY RESULTED IN A COST SAVING OF OVER RS. 250 CRORES.

DRILLSHIPS : EXPLOITATION OF OCEAN RESOURCES

11

S. B. Khalsa,
G. M. (T), ONGC, MRBC, Mumbai.

1. INTRODUCTION

The crude oil production of our country at present is about 32 Mtpa as against the consumption demand of 80 Mtpa. Thus, less than 50 per cent of the demand is met by domestic production while in the late 80s India was producing about 69 per cent of the crude oil demand. However, with the increase in consumption, the gap between demand and production/supply has continuously increased. It is estimated that by the year 2001 the demand will increase to over 100 Mtpa and by 2010 it will reach a level of 164 Mtpa. Although ONGC started the first phase of its offshore exploration in the shallow waters of Gulf of Cambay in ALIABET — 1 in 1962, its offshore activities at Bombay High commenced in early 70s. The production of crude oil on commercial scale started in 1974. At present, the average daily production of oil is 3,42,00 bbls and gas from Bombay High, 49.00 m^3. These fields are mainly in water depths ranging from 30 mtrs. to 90 mtrs.

In order to reduce the increasing gap between supply and demand we must exploit our deep-sea reserves. One of the advanced techniques for this purpose is the use of drillships. This paper outlines the technology available on deep water drilling and the strategy adopted by ONGC.

2. TYPES OF OFFSHORE MOBILE DRILLING RIGS

Offshore rigs can be grouped in two main categories : Jack-ups and Floaters.

(a) *Jack-up Rigs* - This type of rig is provided with fixed 'stable platform and the well head is set at a fixed distance from the rotary table. The well heads and BOPs are usually above the water level and through the jacking system, the rig is moved upward or downward by maintaining the necessary air gap, keeping in view the environmental and technical conditions. This type of rig is good for sea water level above 100 mtrs. The main advantages of jack-up rig are—

- It can work over a platform
- The down-time is less

- Once on location there is no motion
- It is cheaper in cost than floater

However, jack-up rigs have also some limitations which are :

- The rig cannot quickly move off-location in case of blow-out or other mishap.
- Water depth is limited to below 100 mtrs.
- There are chances of punch-through when soil conditions are not tavourable.

Jack-up rigs are of two types

(i) With Independent leg, and
(ii) Mat type

(B) Floaters : Floaters can be grouped into 3 types

(i) Submersible or Barge used for shallow waters. Having two hulls, the lower hull is used for balancing and allowed to rest on the sea-bed.

Advantages:

- Low cost
- Smaller crew needed

Disadvantages:

- Low towing speed

(ii) Semi-submersible floaters

- Used for deep and choppy seas

Advantages:

- Relatively stable drilling platform
- Can operate in deep water

Disadvantages:

- Limited loading capacity
- More support vessels required.

(iii) Drillships

As the name indicates a drillship is a self propelled vessel with all arrangements for drilling an oil well while in floating condition. It is the most mobile of all the three types and has sufficient loading capacity to operate for long periods.

Advantages (i) Deep water capability (ii) Larger loading (iii) Faster travel and (iv) No need for tugs.

Disadvantages (i) Very large heave response due to surface contact with sea (ii) Limited capability to operate in high wind and rough weather which produces excessive platform motion.

To overcome the above problems, drillships are equipped with specialised equipments like heave compensator, riser tentioners pipe racking system, remote operated draw work and pneumatic elevators.

Drillships are also equipped with sub-sea equipments, sub sea BOPs, saturation diving system, sub-sea TV, iron roughnecks etc.

Drillships can be with spread mooring system, dynamic positioning system, turret mooring system with thrusters etc. S/Vijay and S/Bhushan of ONGC are drillships with spread mooring system.

3. DEEP WATER RESOURCES

Based on the preliminary surveys undertaken, ONGC estimates that further 5 to 9 billion tons of oil will be available from deep water areas having 400 to 900 mtr. depth. Because of the increasing gap between supply and demand, deep water drilling has become one of the thrust areas and with the available technology in Europe and America, ONGC decided to go for deep water drilling.

As shallow water oil is depleted, deep water resources have become more important.

Douglas-Westwood's analysis of proposed offshore field shows that 44 per cent of the field's resources are in depths greater than 300 mtrs. Gulf of Mexico and Brazil, deep waters account for 99 per cent of oil and gas reserves, which are considered for development.

At the end of 1996, the production from deep water fields was about 1 million b/day worldwide and is expected to be 3 millions b/day by the year 2000.

A number of factors account for the increase in global deep water activities.

- Depletion of shallow water reserves.
- Deep water has greater potential for large field.
- Improved technology and management practices reducing development cost.

Our understanding of what deep water means changes each year as the offshore industry ventures out into increasing water depths. In the 1970s the term "Deep water drilling" related to depths of 200 and 400 mtrs which has now been revised to between 400 mtrs and 1000 mtrs.

A total of 301 deep water discoveries have been reported world-wide since deep water exploration began. The average size of a deep water field is much larger than that in shallow waters. Success has been achieved in geographically diverse areas like Brazil offshore, the Gulf of Mexico, North Sea, South China Sea and New Zealand.

Enthused by the success stories of Brazil Gulf of Mexico, Norway etc. ONGC has also decided to extend its exploratory activities in the deep waters off India's east west coasts. Preliminary estimates of hydrocarbon reserves in the country's deep water areas up to 1000m bathymetry range between 5 and 9 billion tonnes of O and OEG.

4. THE EXPLORATION FRONT

The recovery of hydrocarbons depends on a number of aspects viz nature of rock, permeability, porosity, type of hydrocarbon etc. In some parts of the world recovery factors have been of the order of 60 per cent or more for sandstone reservoirs, 40-50% for carbonate reservoirs and 25-35 per cent for chalk reservoirs. At present this level of recovery may be achieved under ideal conditions, but recoveries in general, would be 10-15% lower. ONGC has already attempted exploratory drilling in water depths more than 200 mtrs, resulting in two small recoveries, in 1982. This has emboldened ONGCs resolve to explore "high-risk high reward" areas.

ONGC has acquired about 70,000 line km. of seismic data in deep water. Based on seismic and other related data about 5-9 billion tonnes of O and OEG are estimated in deep water. It is estimated that 20-35 per cent of hydrocarbon potential in India is in waters over 200 mtr. deep. The total area of interest is about 1.4 million sq. km. in the Arabian sea and Bay of Bengal and is categorized as under :

A. West Coast deep water offshore basins :

(i) Kutch Saurashtra offshore.
(ii) Bombay offshore.
(iii) Konkan Lakshadweep, Kerala.

B. East Cost deep water offshore basin

(i) Krishna Godavari offshore
(ii) Cambay offshore
(iii) Mahanand offshore
(iv) Bengal offshore
(v) Andaman Nicobar offshore.

ONGC made discoveries of G1 and G2 structures off the East Coast of India in 1982 when six wells were drilled on G1 structures. Out of these four wells were hydrocarbon bearing. Similarly of the three wells drilled on G2 structure, two were found to be hydrocarbon bearing.

Evidence of several prospects on the seismic maps discovery of oil/gas bearing structures in 1982 from G1 and G2 and favourable entrapment conditions have boosted ONGCs confidence to undertake exploration in frontier areas.

5. PRESENT STATUS OF DEEP WATER DRILLING IN INDIA BY ONGC

Accordingly, the first deep water well in India was drilled by ONGC a location CDW-1A(A) to a depth of 2127 mtrs. in Cauvery Basin Offshore (in the water depth of 764 mtrs.) at a cost of Rs. 66.27 crores by the corporations upgraded vessel, S/Vijay. The well was completed by 23-9-98. At present the rig is operating on its second location U-1-A-D in KU Basin, drilling a well with a T.D. of 2500 mtrs (in the water depth of 560 mtrs). The drilling depth achieved sofar is 1000 mtrs.

Drillship "Sagar Vijay" was upgraded in Cochin shipyard during 1997-98. The details of upgradation are as follows :

6. UPGRADATION OF SAGAR VIJAY - DRILLSHIP

A feasibility study for upgrading of the ONGC rig for operating in water depth up to 900 M. was carried out by NSDRC, Visakhapatnam, in the year 1996, who co-opted with MCS international, IHC Gusto Engg. and Moduspec International of the Netherland. Work was sub-contracted to the following parties, each specialized in its respective field.

(A) IHC Gusto Eng. (Gusto) The Netherlands

(i) Station Keeping system
(ii) Modification to hull and structure
(iii) Workability assessment

(B) Moduspec International, The Netherlands

(i) Riser system (Equipment)
(ii) BOP
(iii) Acoustic positioning system in sub-sea TV system.

(C) MCS International, The Netherlands

(i) Riser system (Dynamics)

(D) NSDRC, Vishakhapatnam

(i) Stability calculation
(ii) Operating manual

Upgradations/Modifications recommended vs implemented
The upgradation was carried out in respect of the following :

A. Station Keeping System
B. Purchase Repair of Equipment
C. Rig Modification
D. Hiring of Associated Services

A. *Station Keeping system (Mooring system)*

This includes items, such as, anchors, anchor wires, chains and associated equipment, anchor handing boats etc.

(a) *Anchors* : Number of anchors has been kept at 8 as before.

(b) *Anchor Wires* : In place of wire ropes a combination of wire and chains has been provided to cater for deep water requirements to maintain weight on anchor and reduce tension on anchor winches.

(c) *Chains* : The chains have been taken from rig Sagar Prabhat for use of this rig.

(d) *Associated Equipment* : The chasers and shackles have been purchased and fitted to facilitate anchor handling.

Anchor Handling Boats : the anchor handling boats have been chartered from Ms. Maersk Supple Service, having main engine capacity of 12,000 bhp, winch drum capacity of 4 × 1050 mtrs. × 76 mm and J-type chain chaser and grappling hook for 1100 mtr. water depth.

B. *Purchase/Repair of Equipment*

This includes purchase and repair of BOP and Control System LMRP, Riser Package and associated equipment, Riser Tensioner System, Buoyancy Module for Risers and Acoustic Positioning System.

(a) BOP and control system

This includes retrieval of old BOP, dismantling, inspection and repair of same.

(i) Status of BOP and Control Unit.

- This has been attended by OEM M/S Shaffer Singapore.
- The control unit was also serviced.
- Four additional accumulator bottles installed.
- APV bottles dismantled from S/Prabhat to be used.
- New Pod hosed purchased and made available,
- Pod hose reels overhauled.

(ii) Status of LMRP

- LMRP lost during cyclone in June 1996
- Procured from M/s. N.L. Shaffer, the OEM

(iii) Since Sub-sea services are not possible in deep water drilling, ROV services are hired for sub-sea operations.

(b) Riser Packing and Associated Equipment

- Riser Package includes Flex joint with Riser Adapter, Mud Booster termination Spool, Drilling risers. Riser auto fill up valve and Telescopic joint.
- Status-Diverter flex Joint , Riser handling and tools like Spider Gimbal and Running tool and Support ring designed, modified and provided.

(c) Riser Tensioner System

- Four additional tensioners taken from S/Prabhat and installed at appropriate positions.
- Modifications made to accommodate Riser Tensioners in position.
- Installed 7 Nos. additional Air Pressure Vessels and 4 Nos. high Pressure Compressors to enhance capacity of Riser Tensioner and Motion Compensator Air Bank.
- Replaced all winches on derrick and Moon Pool areas with higher capacity.
- Installed 4Nos new Sub-sea Air Compressors of 2850 PSI, Model 5417, Make-M/s. COMPARE REVEL
- Removed the Guide and POD line System from the Rig S/Prabhat and Kept as spare.

(d) Buoyancy Module for Risers :

- New Buoyancy module purchased and fitted on rig.

(e) Acoustic Positioning System :

- The existing Simrad Acoustic Positioning System has been upgraded by installing narrow beam transducer.

C Rig Modifications :

This includes structural work for installing additional tensioners Air pressure vessels and high pressure air compressors and control panel for tensioners. This also includes installation of a third on mud pump, closing of diving Moon-pool and cutting of bulkheads and installation of new SCR systems. Details of these modifications are as follows :

- Additional space has been created by removing the old unusable equipment mainly for diving operations. The associated modification to store 70 feet riser with buoyancy module and deck strengthening was carried out.

- The third mud pump 12-P-160 which was initially diesel driven (mechanical drive) and taken from ERBC has been converted for electrical operation. The SCR system from Hill-Graham and MCC panel from the Cochin shipyard has been procured and fitted.
- The old Shale Shaker replaced with linear motion Shale Shakers, new Scalper systems installed for Gumbo problem, old mud mixing system replaced with jet shearing system, old analog anchor line measuring system replaced with digital display along with strain tension meter.
- The old drilling instrumentation was replaced with new one computerized based LCO type, having mud watch digital display and capacity to handle the third mud pump.

D. Hiring of Associated Services

The Remote Operated Vehicles (ROVs) and anchor handling boat services have been hired from outside agencies.

7. CHALLENGES IN DEEP WATER DRILLING AND TBG ROLE

(i) *Anchor handling* : Because of the restrictions of divers going into the deep sea it is not possible to get the soil properties and other parameters of the sea-bed accurately. As a result it is very difficult to select suitable type of anchors and we have to use hit and miss methods to engage the anchors and provide stability to the drillship. Certainly there will be more load and more wear and tear on the anchor winches and hence more attention has to be given for their proper functioning. Strain and tensiometers provided within the system are to be properly calibrated and monitored. The lowering of the chain with the boats and then connecting it to the anchor of the ship is going to be critical operation. In case of harsh environment special care is needed to be taken.

Dynamic positioning is the latest advance in securing a vessel without anchoring. In emergency conditions such as extreme weather, in or in case of operational problems the ship can disconnect itself from the well in less than a minutes and can move away from the location. There are four types of deep water spread mooring system as shown in Figs. 11.1 to 11.4.

(ii) *Mud engineering and handling equipment* : Due to low temperature on sea-bed when mud is coming from the well it is necessary for homogeneous and adequately conditioned mud to encounter with the high pressure of the well, as well as to control the loss of the well we have to use the advanced technology and latest developed mud handling equipments like linear motion shale shaker, mud cleaner, mud shearing mud mixing pumps and meet the requirement of the mud to use the third mud pump.

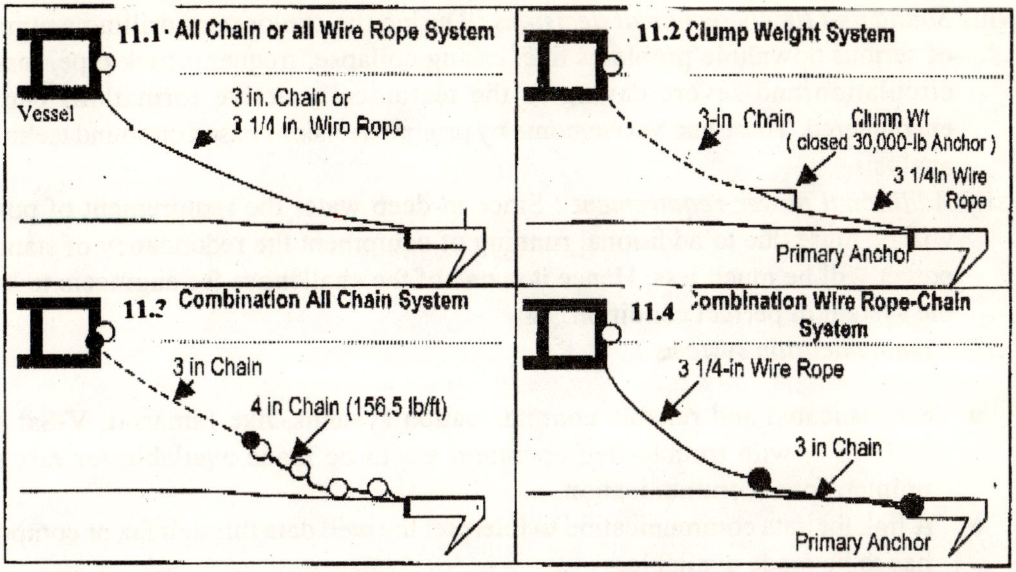

(iii) *Instrumentation* : Each individual system equipment needs to be provided with sophisticated instruments and gauges for safe and reliable operation including Driller-Console. The analog type gauges are to be replaced with digital ones to have more reliable measurement. The instruments should function correctly even in harsh environment for taking correct decisions. TBG's role to repair and maintain these instruments. Sufficient number as standby need to be provided. Suitable system is to be introduced for repair and calibration of gauges and prompt action needs to be taken for repair/replacement of defective sets.

(iv) *Low temperature on the sea bed* : Because of low temperature and high water pressure on the sea-bed it is not possible to attend the operations at sea-bed with the help of divers. The use of latest technology like ROV, diverless operations, guidelineless operations etc. is essential for reducing downtime and optimum utilization of costly resources.

(v) *Risers* : For drilling in deeper waters riser is a vital link between the surface and sub-sea system. The design of the riser is effected by many factors, such as buoyancy, tension to be applied, current depth at location, wave loading weather etc.

(vi) *Sub sea control system* : In the absence of divers, a closed circuit television system is used to guide operations in deep water. The well control devices like blowout preventor is placed on sea-bed with the help of ROVs.

(vii) *Current* : In deeper water, drilling has its own special operational problems. It requires proper technical pre-planning as the currents in deep water are stronger. The strong surface and under water currents impose additional high stresses on the riser. The re-entry operation also becomes complicated as the riser is subjected to current forces from various directions.

(viii) *Sound and timely technical decisions* : During the exploratory drilling a number of serious downhole problems like, casing collapse, frequent stuck pipe, loss of circulation and severe caving in the tectonically active formations can be encountered. These can be overcome by prompt decisions based on sound technical analysis.

(xi) *Additional power requirement* : Since in deep water the requirement of power will be more due to additional running of equipment the redundancy of standby power will be much less. Hence it is one of the challenges for engineers to keep the system in perfect condition.

(x) *Communication systems :*

- Sophisticated and reliable communication systems, like Inmarsat, V-Sat and VHF/ISV, with trouble-free operation are to be made available for fast and uninterrupted communication.
- A link for data communication to interpret the well data through fax or computer has to be made available.

In addition to the above the technical aspects which play a significant role in the critical areas are as under :

a. The selection of a suitable drillship after studying the detailed specifications, design etc. which should be techno-commercially viable and suitable for Indian operating conditions and environment.

b. The selection of equipment for better performance and control :

- Provision of linear motion shale shaker in place of the old system.
- Use of mud cleaner of better conditioning of the mud.
- Use of top drive for fast operation.
- Use of remote operated Iron Roughneck.
- Use of mud shear mud mixing pumps.
- Pipe handling system.
- Station keeping systems.

c. Coordination with statutory bodies, like ABS. MMD and Ministry of Environment and Forests for meeting statutory requirements.

d. Before positioning of the rig the collection of meteorological data from local as well as international weather bureaus of repute for accurate weather prediction.

e. To co-ordinate with the various logistic and chartering agencies for providing necessary assistance during rig move preparation and start of the operation immediately after the rig is positioned.

f. To maintain the good health of the equipment, a regular condition monitoring of the equipment with the advanced sophisticated testing equipment and also to follow the standard of API religiously for repairs and maintenance.
g. A training course designed for technical officers to be conducted at regular intervals so that they update their knowledge.
h. In order to meet any emergency during the operation the Engineering Superintendent to be given special authority for procurement of urgent items.
i. If the results of deep water drilling are encouraging the major role ahead for the TBG will be in the following area :

- The deep water wells will require sub-sea well completion and these connected to floating process platforms as against the existing fixed structure platform. Hence design, fabrication and installation will be TBG role.
- The laying of pipelines in sea water and the methodology for their support will need study before the job undertaken. The repair thereafter is going to be a challenge to TBG, which will need a team of trained engineers.

CONCLUSION

For ONGC Deep Water Exploration is a virgin area also a "high risk/high reward" venture. The foremost pre-requisite 'finance' is not really a serious problem a as for technology for deep water drilling which ONGC is looking, it is globally available.

ONGCs current plan is to drill four wells costing approximately Rs. 250 crores. To minimise the risk, the project will be spread over. In view of the keen interest shown by certain cash rich private interests and public sector oil companies, ONGC is examining possibilities for joint venture.

IV

MAN POWER AND TRAINING

AUTOMATIC GENERATION OF ALL POSSIBLE COMBINATIONS OF DAMAGED COMPARTMENTS FOR PROBABILISTIC DAMAGE STABILITY CALCULATIONS

12

Tanmay Sarkar, D.G. Sarangdhar
Research and Coordination Division
Indian Register of Shipping, Mumbai-400072

ABSTRACT

Probabilistic damage stability rules are introduced as IMO Resolution MSC. 19(58) and also included in SOLAS conventions as Part B-1 of Chapter II-1 to absorb probabilistic nature into the prediction of survivability and safety. In these rules available stability index and required stability index are calculated and compared. To find these indices one has to generate all possible combinations of compartments in a ship taking one, two, three etc. of them together. For a complex design this forms an enormous task when done manually.

This paper presents an efficient technique for generation of all possible combinations of damaged compartments taken one, two, three etc. at a time for use in probabilistic damage stability calculations of cargo ships. The method is based on a branching technique of finding the combinations. As inputs one has to supply only the end co-ordinates of the compartments and list of its neighbours. These combinations can then be used as a input to the damaged stability program available in IRS, to find GZ, range, angle of heel for use the calculation of above indices. The technique is applied to a coastal tanker. Altogether 68 possible combinations of compartments are found for this vessel. On calculating it was found that available stability index 'A' is greater than required stability index 'R'. Whole calculation has taken a couple of minutes which otherwise would have taken a number of days.

INTRODUCTION

It is very important to ensure safety of a ship both in intact and damaged conditions. Damage stability rules are formulated for achieving safe ships, which can withstand certain amount of damage. Rules for damage stability of ships can be categorized into two: (a) Survivability after damage (b) Safety or relative safety of ships. If a ship can survive heavier accident or large damage size it is safer. Damage stability rules are of two types: (1) Deterministic and (2) Probabilistic.

Deterministic rules for damage stability of ships are in Chapter II-1 Part B of SOLAS. These rules apply to both passenger and cargo ships. Probabilistic damage stability rules cater to the probabilistic nature of the survivability and safety. Survivability depends on (a) probability of occurrence of damage and (b) chances of surviving these damages. Probability of survival of damage will depend on the location and extent of damage and environmental conditions. These are generally obtained by analyzing damage data statistics.

Probabilistic damage stability rules are at present applicable to cargo ships including RO RO ships as par SOLAS Chapter II-1 Part B-1[1,3] or IMO Resolution MSC 19(58) and for Passenger Ships IMO Resolution A. 265. Provisions of A.265 [2] need not be applied if the deterministic rules in SOLAS Chapter II-1 Part B is complied with.

Details of probabilistic damage stability rules are explained in SOLAS Chapter II-1 Part B-1, regulations 25-1 through 25-10 [1,3]. Here, these rules will be mentioned only briefly.

These regulations are intended to provide ships with a minimum standard of subdivision and require that attained subdivision index for a vessel should not be less than required subdivision index.

Subdivision Index: The degree of subdivision to be provided shall be determined by the required subdivision index R, as follows:

$$R = (0.002 + 0.0009\ L_s)^{\frac{1}{3}} \qquad (1a)$$

or,

$$R = 1 - [\frac{1}{(1 + \frac{L_s}{100}\frac{R_0}{R_0})}] \qquad (1b)$$

Where, R_0 is R calculated according to (1a).

The attained subdivision index A, calculated in accordance with this regulation, shall not be less than the required subdivision index R.

The attained subdivision index A shall be calculated for the ship by the following formula:

$$A = \sum p_i s_i \qquad (2)$$

where:

i represents each compartment or group of compartments under consideration

p_i accounts for the probability that only the compartment or group of compartments under consideration may be flooded, disregarding any horizontal subdivision

s_i accounts for the probability of survival after flooding the compartment or group of compartments under consideration, including the effects of any horizontal subdivision.

The calculation of attained subdivision index involves identifying all possible group of compartments which may be flooded together. Compartments may be taken in a group of 1,2 .... ,*n*, where *n* is the maximum number of compartment. Limiting factors are the maximum non-dimensional damage length, which should not be more than 0.24 and compartments taken as a group should be adjacent.

To find all possible combinations of compartments is a tedious task. If done manually this may take a number of days. Also there is a possibility that some of the possible combinations may be overlooked. To overcome these difficulties in this paper we present a computer-based technique of generating these combinations. The method uses branching technique and generates all possible groups of compartments within a couple of minutes. Details of this method are as explained below.

1. Method

Identify each compartment by an identification number (say 0 to n-1)
For each compartment take as input its list of neighbouring compartments.

2. Generating combinations of two compartments

This is done by taking one of the compartments (first compartment) at a time. Then for each such first compartment, its neighbouring compartments are selected one at a time as the second compartment.

3. Generating combination of three compartments

(a) For this again we take one compartment (first compartment) at a time.

Then for each first compartment its neighbouring compartments are selected one at a time as second compartment

And again for each such second compartment its neighbouring compartments are selected one at a time as third compartment.

Or

(b) Take one of the compartments (first compartment) at a time and for each first compartment take all possible combinations of its neighbouring compartments taken two at a time as second and third compartments.

4. Generating combination of four compartments

(a) For this take one compartment (first compartment) at a time.

Then for each first compartment its neighbouring compartments are selected one at a time as the second compartment.

Then for each second compartment its neighbouring compartments are selected one at a time as the third compartment.

And for each such third compartment its neighbouring compartments are selected one at a time as the fourth compartment.

Or

(b) Again take each compartment one at a time (first compartment)

For each first compartment its neighbouring compartments are selected one at a time as the second compartment.

For each second compartment take all possible combinations of its neighbouring compartments taken two at a time as the third and fourth compartment.

Or

(c) Again take each compartment one at a time (first compartment)

Take for each first compartment all possible combinations of its neighbouring compartments taken two at a time as second and third compartments.

Select one compartment at a time from the neighbour list of these two compartments as the fourth compartment.

Or

(d) Select one of the compartments at a time (first compartment)

And select all possible combinations of its neighbours taken three at a time as the second, third and fourth compartments.

Similarly all possible combinations of five or more compartments can be found. This program at present can cater up to a maximum of six compartments taken at a time. However, it can easily be extended to more than six compartments. Fig. 12.1 below illustrates the technique by a simple example when 1st compartment is 0.

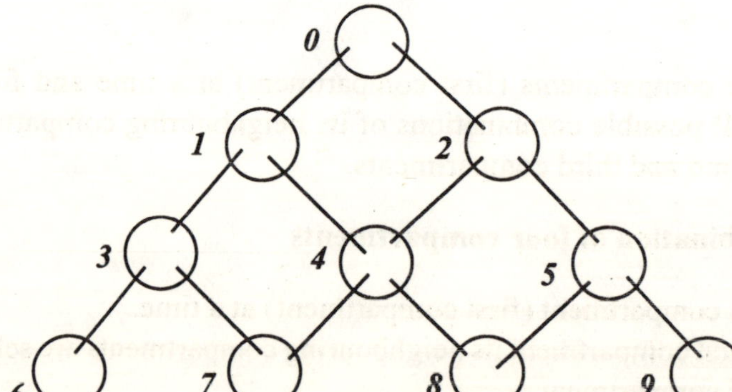

Fig. 12.1

In the above Figure
Neighbours of 0 th compartment are 1 and 2
Neighbours of compartment 1 are 3 and 4
Neighbours of compartment 2 are 4 and 5
Neighbours of compartment 3 are 6 and 7
Neighbours of compartment 4 are 7 and 8
Neighbours of compartment 5 are 8 and 9

Following are combinations which can be generated using 0 as the first compartment by using techniques 2, 3(a) ... 4(d).

2	01, 02
3(a)	013, 014, 024, 025
3(b)	012
4(a) ...	0136, 0137
	0147, 0148
	0259, 0258
	0248, 0247
4(b)	0134, 0245
4(C)	0123, 0124, 0125

Similarly, taking other compartments as first compartment all possible combinations can be generated.

Elimination of the combinations which are repeated

Finding combinations this way will result in repetitions. These repetitions are to be eliminated. The elimination is done in the following way:

Group the combinations according to the total number of compartments. Take any member of the group and compare with all other members of the group. If for any two members of the group matching is found for one compartment they are compared for second matching, third matching and so on. If all the compartments making the two members match one of them is eliminated. In these way all repeated combinations can be removed.

From the remaining, all those group of compartments for which non-dimensional length between after most and fore most point is greater than J_{max} ($48/L_S$, but not more than 0.24) are eliminated. What remain after this elimination, present all possible combinations of the compartments to be considered for probabilistic damage stability calculations.

Once the possible combinations are found attained stability index is given by:

$$A = \sum p_i s_i$$

$p_i s$ for each combination is calculated following the steps details of which are given in Regulation 25-5 of SOLAS Chapter II-1 Part B-1 [1, 3].

Calculation $s_i s$ are done according to the steps given in Regulation 25-6 of SOLAS Chapter II-1 Part B-1[1, 3]. These calculations require information regarding GZ_{max}, range and $_e$, the definitions of which are given below.

Range: range of positive righting levers beyond the angle of equilibrium (in degrees) but not more than 20 degrees; however the range shall be terminated at the angle where openings not capable of being closed watertight are immersed.

GZ_{max}, that is maximum positive righting lever (in meters) within the range as given below but not more than 0.1m

q_e = final equilibrium angle of heel (in degrees)

These quantities are obtained intact and damage stability program of Indian Register of Shipping (IRS).

Once p_i and $s_i s$ for all possible combinations are calculated the attained subdivision index can be calculated using formula (2) and compared with required subdivision index. A Flow chart of the method is given below.

Flow Chart

Calculate Required subdivision index

↓

Start Calculation of p_i & s_i

↓

Find Out All Possible Combinations of adjacent compartments Taken one/two/more at a Time which Contribute to Attained Subdivision Index

↓

Calculate p_i & $s_i s$ for all such

↓

Check whether attained stability index > required index

Results:

To demonstrate the technique it is used for a coastal tanker. Main particulars of the vessel are given below.:

Length (P.P)	89.950	Breadth (MLD)	17.500
Depth (MLD)	8.410	Draft	7.490
Deadweight	7200.0	Gross Tonnage	3950.0
Speed Service	12.500	Speed Max	13.000
Number of Compartments	13		

Using the present technique initially a total number of 92 combinations of compartments were generated. Out of which 24 are eliminated and finally 68 combinations are left for consideration for available stability index calculation.

Using these combinations attained subdivision index is calculated. Range and GZ_{max} values required for these calculations are obtained using damage stability program of IRS. It was found that attained subdivision index for this vessel is 0.443555 against a calculated value of required subdivision index of 0.187888. Whole calculation took only half an hour of time. If done manually, this would have taken more than a man-week. This clearly shows the relative advantage of the present technique in probabilistic damage stability calculations

CONCLUSIONS

A computer-aided technique has been developed for finding all possible groups of compartments of a ship required in calculation of attained stability index of a ship. It is shown by an example of a coastal tanker that the method is efficient. It reduces required time and effort considerably. Thus can be used for evaluation of probabilistic damage stability.

ACKNOWLEDGEMENT

The authors wish to thank Chief Executive Mr. K.Madhok and Chef Surveyor(TS) Mr. A.R. Chitnis for providing necessary support and encouragement to carry out this study and publish the results. Thanks also due to Mr. Rahul Paul, final year B.Tech. student of Ocean Engineering and Naval Architecture Department of Indian Institute of Technology, Kharagpur for carrying most of numerical calculations repeated in this paper.

REFERENCES

1. International Convention for the Safety of Life at Sea, 1990/1991 Amendments
2. IMO Resolution A.265 (VIII): Regulations on Subdivision and Stability of Passenger Ships as an Equivalent to Part B of Chapter II of the International Convention for the Safety of Life at Sea 1960, International Maritime Organization, London (Adopted on 25 May 1990)
3. IMO: SOLAS Consolidated Edition, International Maritime Organization, London, 1992

STEELS FOR MARINE APPLICATION: PROPERTIES, PROCESSING & FABRICATION

13

KN Pandey
R & D Centre for Iron & Steel,
SAIL, Ranchi

ABSTRACT

Selection of steel for marine application whether off shore structure, jackets, vessels or pipe lines etc. is based on environmental conditions, functional loads and method of fabrication deployed. Apart from functional loads, steels in marine application are subjected to cyclic loading , low temperature, corrosive atmosphere, which cause sharp decrease in fatigue strength of steel and increased fatigue crack growth rate in welds leading to premature failure. These stringent requirements demand ultra clean high purity product with requisite chemistry and treatment,which combine the properties of excellent weldability, resistance to corrosion induced cracking and improved impact toughness.

To meet these requirements, many developments like supply of pretreated hot metal from closely controlled blast furnaces, steel making through basic oxygen furnaces, wide spread introduction and use of ladle refining processes, argon stirring, continuous casters with electromagnetic stirring and air mist cooling of strand, closely controlled rolling, thermo mechanical-treatment etc. has taken place.

Different pre-welding and post-welding techniques, like use of special electrodes and improved welding to increase heat affected zone (HAZ) toughness and decrease HAZ hardness, TIG dressing plasma dressing, weld toe grinding, shot peening etc. are employed to achieve required weld quality. In the present paper, quality requirements of steel for marine application, processing of the steel for achieving required quality and latest fabrication processes have been discussed. Apart from the above, the status of indigenous availability of steel for marine applications and its processing in SAIL plants has been also mentioned with description of facilities available and process route followed.

1. INTRODUCTION

Steels in marine application are subjected to corrosive atmosphere, cyclic loading and in many cases extreme variation in temperature, apart from the functional loading required. The fatigue strength ratio in freely corroding sea water environment is drastically reduced as compared to that in air, as shown in Fig. 13.1 .These additional adverse factors cause failure of the steel much below the expected stress levels. Many developments have taken place to take care of these problems in the area of introducing

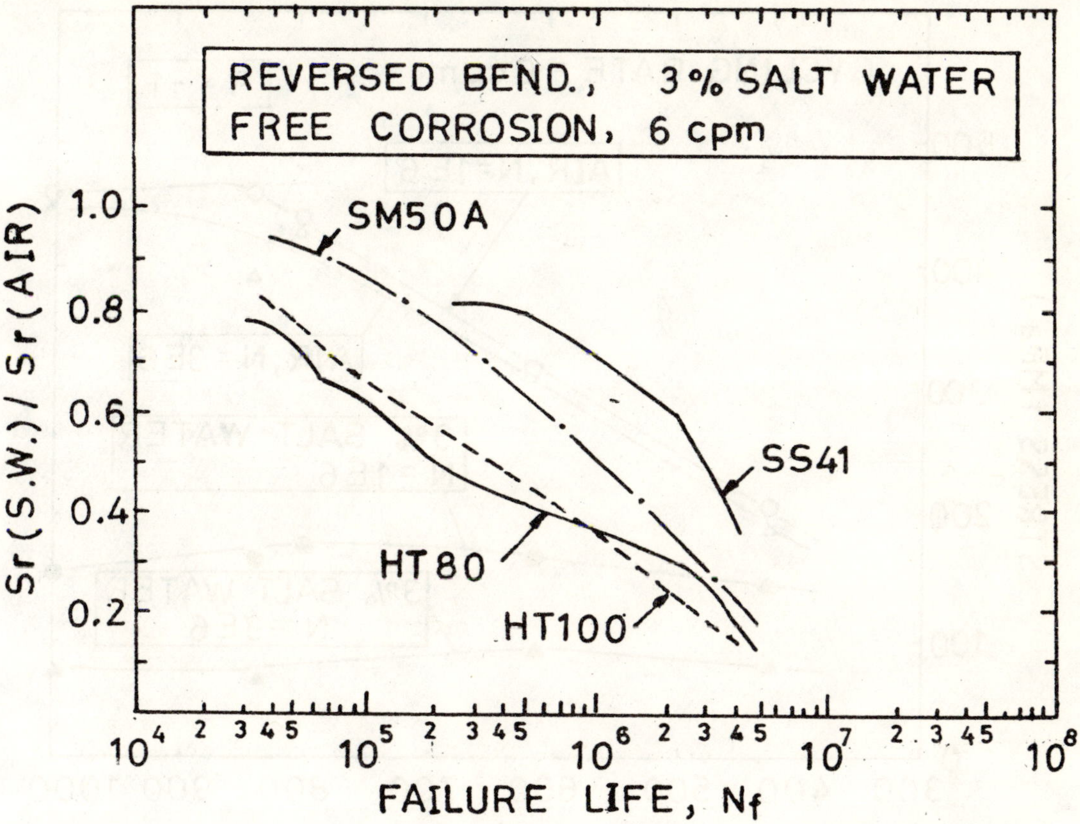

Fig. 13.1 : Ratio of fatigue strength in salt water to that in air as a function of failure like.

new products with high strength, but more than often this does not solve the problem, as only high strength is not sufficient to prevent failure and impart long life. This is because the fatigue strength of smooth specimen tested in air increases with increasing tensile strength upto a certain level and after that shows a saturated behaviour. On the other hand, the fatigue strength in salt water holds almost constant value regardless of tensile strength, as shown in Fig. 13.2. Hence, properties like cleanliness of steel, weldability, toughness of weldment, microstructure of heat affected zone, pre-welding and post-welding processes play a decisive role in deciding the overall performance of product.

2. PROPERTIES

Steels used for marine application must have high resistance to fatigue, fracture, and corrosion, as these are subjected to wind, wave, current and temperature variation induced strains. But these requirements are not sufficient, e.g. weldments should have high resistance to corrosion induced crack generation. To achieve better weldability, drastic reduction in carbon (0.2 to 0.1%), and improving resistance to lamellar tearing

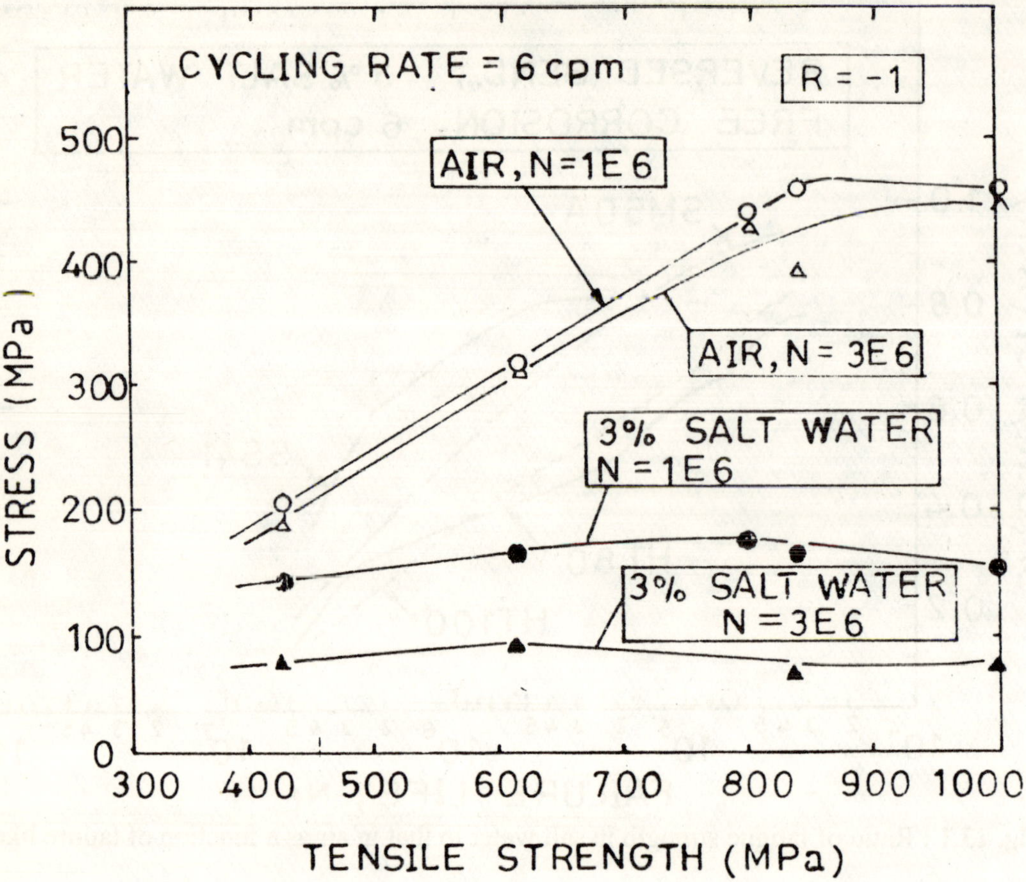

Fig. 13.2 : Fatigue strength as a function of static tensile strength.

and impact toughness, decrease in sulphur content (0.04 to 0.01% or below) is required. Presence of residual elements should be minimum. Otherwise they can cause embrittlement by segregating to grain boundaries. If it occurs there is a marked increase in the tendency towards intergranular fracture along the austenite grain boundaries. This represents a potentially serious reduction in fracture resistance that can also increase the risk of stress corrosion fatigue and stress relief cracking. Embrittlement is most often caused by segregation of elements such as Br, Sb, Sn, As and P. The likelihood of a critical degree of embrittlement being caused is a very complex function of steel type, grain size, grain boundary cohesion and matrix strength.

The risk of embrittlement is kept low and in balance by the careful use of other alloying elements such as Mo, Ti and rare earth metals. These elements reduce the solubility of the embrittling element by precipitation of the compounds. Brittle fracture possibility increases with lower temperature and high stresses especially if there are three axial stresses in a thick plate. Stress relieving gives more safety but the main requirement is to have a material with good toughness and relatively large allowable defect size. Adequate toughness to guard against the initiation of fast fracture is also to be ensured.

Weldability is another important property which needs serious consideration so that the risk of fracture initiation within the coarse grain region of heat affected zone adjacent to the fusion line is avoided. Other important factors to be taken care, of, are strain ageing characteristics, cold cracking, hydrogen cracking and susceptibility to heat affected zone hazard. To meet the above requirements, steels must be low in carbon and ultra clean, i.e. with least impurities. In order to limit the maximum hardness and risk of cold or hydrogen cracking in the heat affected zones of welded joints, off shore specifications are now requiring much lower carbon ($C \leq 0.12$ and $C_{eq} \leq 0.40\%$) and phosphorus content below 0.010% which improves the toughness of weld joints. Sulphur content of 0.008% or even 0.005% is desirable for avoiding centre line segregation and getting uniform mechanical mechanical properties through out the thickness. Reduced hydrogen content of liquid steel ($\leq$ 2ppm) secures high internal soundness. Low levels of oxygen ($\leq$ 10 ppm) and nitrogen ($\leq$ 0.012%) give desirable properties.

3. PROCESSING OF STEEL

3.1 Steelmaking & Continuous Casting

Bulk steel production in India is through blast furnace, basic oxygen furnace and continuous casting route. Electric steel making either through electric arc furnaces or induction furnaces caters to low alloyed or stainless steel requirements. Hot metal from closely controlled blast furnaces is desulphurised and taken to basic oxygen furnaces, commonly known as LD converters. Argon is purged in the liquid steel for the chemistry and temperature homogenisation. If the requirement demands ladles are taken to vacuum arc degassing (VAD) and vacuum oxygnen decarburisation (VOD) units, for removal of entrapped gases and alloying in VAD and for decreasing carbon in VOD units. Liquid steel ladles are brought to continuous casting machines for casting.

Continuous Casting

Continuous casting of steel has resulted in marked improvement in respect of cleanliness and surface and internal quality of semi-manufactured product required for rolling. Product quality in continuous casting has attained high level, in general, exceeding the equivalent ingot route in most cases by virtue of better surface quality and more uniform internal soundness in particular. Quality depends on three factors, viz. steel cleanliness, surface quality and inner soundness.

Steel Cleanliness

Steel cleanliness depends to a large extent on steel making practice but is very much affected by the mode of casting operation as well as reoxidation of molten steel due to air aspiration into the ladle stream and slag contamination. Slag from ladle to tundish is increasingly prevented with the help of electro magnetic slag detector. The tundish level must be protected with a slag of low oxygen potential assured by highly basic composition. In the mould, steel level should remain constant. The top slag formed from mould powder can be entrapped by turbulent steel flow. Electromagnetic stirring (EMS) in the mould with counter flow appears to provide effective turbulence control in slab casting, apart from improving the micro cleanliness in general by stream protection. For the floatation of smaller inclusions, i.e. below 50 microns in diameter, argon rinsing is done in tundish bottom. Since inclusion frequency is highest at the start of the casting and during ladle draining, raising of low steel temperature during such period by tundish heating also improves steel cleanliness.

Surface quality

Surface quality essentially depends on solidification control of the initial shell formed at the meniscus in the mould. Surface quality requirements are even more stringent in direct rolling with no defect removal from the surface and the stainless steel processed without grinding. High casting speed favors quality. Invariably, most surface defects are found during regimes of low casting speed, because of reduced convective flow in the mould which results in uneven shell growth and surface accumulation of slag and pinholes. Accelerated flow (about 0.5 m/sec.) by means of EMS in the mould is now often used as counter measure in billet and bloom casting. A general prerequisite for good surface quality is the prevention of irregular casting due to nozzle clogging, viscous mould powder and mould level fluctuations.

Inner soundness

Inner soundness demands prevention of microsegregation connected with interdendritic cracking and centre line shrinkage. The former is still a major problem with steels that form low melting constituents (sulphides, phosphides, borides), requiring limit on the casting speed to maintain the mechanical stress level in the strand shell strength. A different apporach is pursued by extremely strong secondary cooling improves inner soundness on account of reduced solidification time and a final solidification structure.

3.2 Rolling

A few decades ago, rolling was considered a simple process for imparting the required shape and size to the final product. The product was normalised for relieving the stresses produced due to rolling, refining the micro-structure and achieving the yield strength of thick plates in the order of 400 MPa. Presently rolling stands with high rigidity, increased drive power and automatic gauge control are used to satisfy the strict tolerance requirements. High rolling forces and torques are required for rolling at low temperature and large thickness reductions. Alloy steels offering a higher deformation resistance, represent an additional difficulty. Now a days, essentially three different techniques are used to roll plates.

- Conventional hot rolling
- Normalising rolling
- Thermo-mechanical rolling

In normalising rolling, the time temperature profile during the deformation or reduction is controlled in such a way that final reduction takes place in the normalising temperature range. The fine grained structure that is achieved, makes it possible to do without any subsequent annealing (normalising). The particular advantage of this technique, apart from energy saving, is better surface quality of rolled material.

Thermo-mechanical rolling, is the centre piece of the rolling process development that is undertaken in conjunction with materials developed. It is closely linked with the development and use of microalloyed steels. With this technique, the final reduction takes place in a low temperature range in which the deformed microstructure does not undergo recrystallisation prior to its transformation. The transformation structure emerging from such strengthened austenite leads to good combination of properties. Plates of higher strength, combining good toughness are produced by means of quenching and tempering (reheat quencing or direct quenching). Effect of different carbon equivalents and process on yield strength has been shown in Fig. 13.3.

Many plate mills are equipped with cooling facilities in order to cool the rolled plates in line directly after rolling in a continuous process, at an accelerated rate

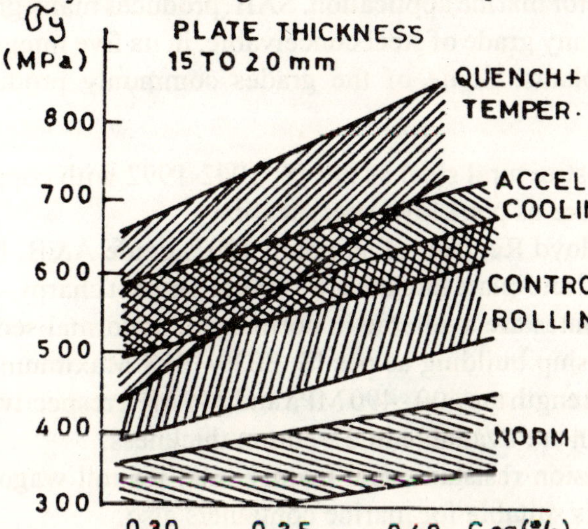

$$C_{eq} = C + \frac{MN}{5} + \frac{Cr + Mo + V}{5} + \frac{Ni + Cu}{35}$$

Fig. 13.3 : Evolution of Oy VS. C. E. for various grades of structural steels.

(accelerated cooling). The cooling rate achieved is around 15°C per second. The rolled material produced in this way has improved microstructure and properties. Controlled rolling allows achievement of the desired mechanical properties directly at the outlet of the rolling mill by on line control of the product temperature and of the rolling forces. Guaranteed yield strength of 600 MPa (or even more) is achieved with CE < 0.45. on plate thickness in orde to get finer grain structure. Accelerated cooling is performed after the final control rolling sequence. This allows either improvement of the mechanical properties of the plate for grain analysis or for reduction of alloying elements for given grade. This method is applied for thickness higher than that can be attained by controlled rolling. The quenching and tempering of the plate is designed to produce microstructures different from those through processes described above. Elements such as C, Mn, Cr, Ni, Mo etc. are added to produce martensitic structure after quenching.

4. STEELS PRODUCED IN SAIL PLANTS

Steel Authority of India Limited (SAIL) produces many grades of quality steel for different applications to the exact national and international specifications as per the stringent requirements. They are produced in plants having all the facilities required for manufacturing these grades. SAIL plants are producing quality hot metal from closely controlled blast furnaces, liquid steel from basic oxygen furnaces/electric arc furnaces. They are equipped with the latest continuous casters, rolling mills and facilities for different treatments. A typical process flow for the complete manufacturing cycle has been shown in Fig. 13.4. Many grades and sizes of rolled sections and flat products (Plate, strip) are manufactured for marine application. SAIL produces many grades of steel and is capable of producing any grade of steel conceivable, in its five integrated steel plants and two alloy steels plants. Some of the grades commonly produced for marine application are :

- Corrosion resistant structural steel as per IS: 2062-1992 with copper content.
- Steel for hull structure as per Lloyd specification
- Steel as per SBQ Lloyd Register of shipping 1991, grade A&B, UTS-400-490 N/mm^2, YS 235 N/mm^2 and plate thickness 8 - 40mm. Impact charpy - notch toughness at 0°C 27J. These plates are available in normalised/unnormalised condition.
- Structural steel for ship building as per IS: 3039-1988 Maximum ultimate tensile strength and yield strength are 400 - 490 MPa and 235 MPa respectively. These plates with 22% elongation are available in 25-50mm thickness.
- SAIL-COR - Corrosion resistant steel mainly used for rail wagons, coaches and armoured vehicles is suitable for marine containers also.
- SAIL-SAGAR - This high strength low alloy steel has been developed for Indian Navy. This steel combines excellent properties of high strength, toughness and weldability.

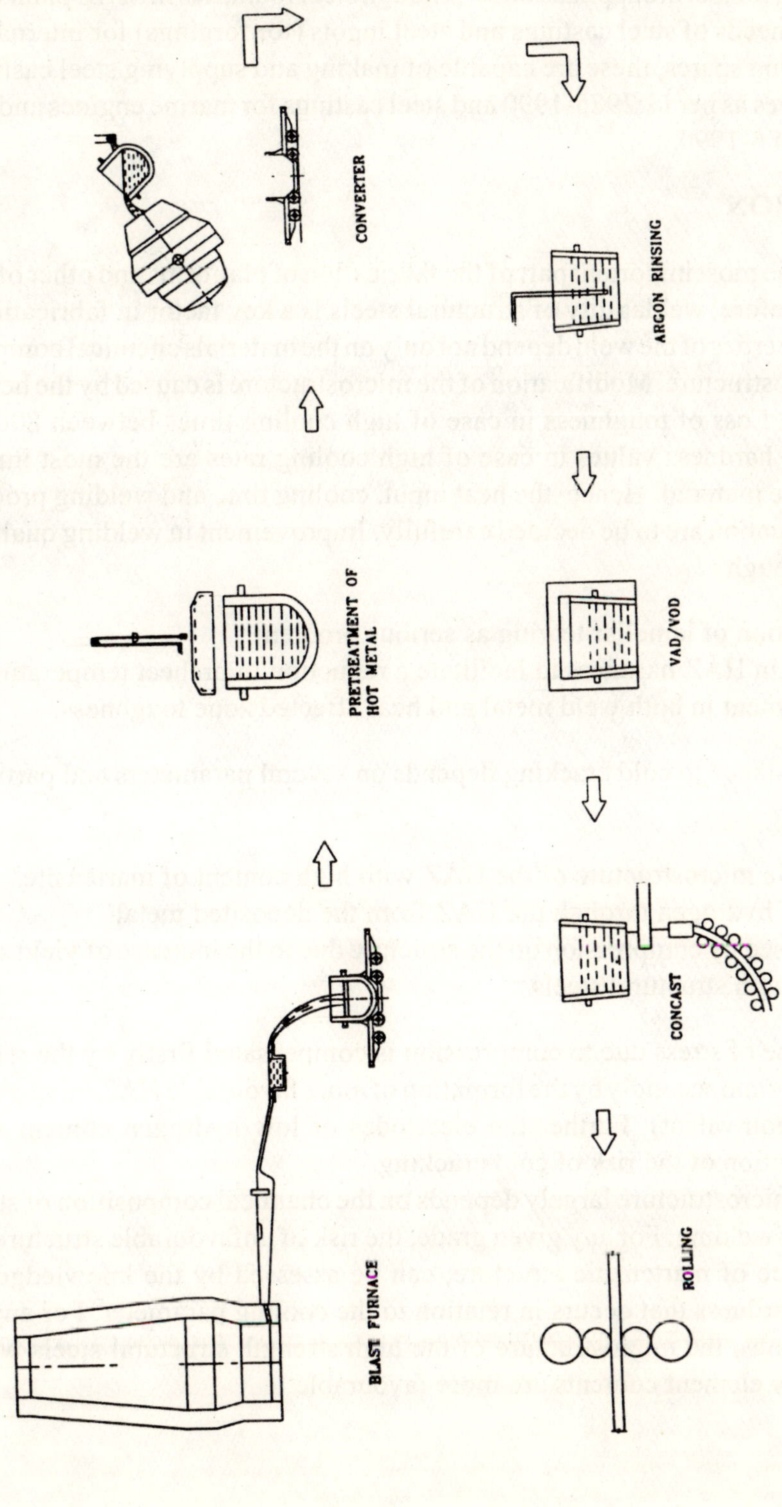

Fig. 13.4 : Process route of steel making.

- Steel casting for marine applications—Though steel foundries in SAIL plants mostly cater to the needs of steel castings and steel ingots (for forgings) for internal use in manufacturing spares, these are capable of making and supplying steel castings for ship structures as per IS:2985-1990 and steel castings for marine engines and boilers as per IS:2986-1990.

5. FABRICATION

Weldability is the most important part of the fabrication of platforms and other off-shore structures. Therefore, weldability of structural steels is a key factor in fabrication. The mechanical properties of the weld depend not only on the materials chemical composition but also its microstructure. Modification of the microstructure is caused by the heat input during welding. Loss of toughness in case of high cooling times between 800°C and 500°C and high hardness values in case of high cooling rates are the most important influences on the material. Hence, the heat input, cooling time and welding procedures used in the fabrication are to be decided carefully. Improvement in welding quality may be achieved through :

- The elimination of lamellar tearing as serious problem.
- A reduction in HAZ hardness to facilitate a reduction in preheat temperature.
- An improvement in both weld metal and heat affected zone toughness.

The lack of resistance to cold cracking depends on several parameters and particularly on :

- Unfavourable microstructure of the HAZ with high content of martensite.
- Diffusion of hydrogen through the HAZ from the deposited metal.
- Stresses caused by compression on the structure due to the increase of yield strength of high strength structural steels.

The increase of stress due to compression is compensated firstly by the reduction of plate thickness and secondly by the formation of more favourable HAZ microstructure (lower carbon equivalent). Further the electrodes of low hydrogen content allow a significant reduction of the risk of cold cracking.

The HAZ microstructure largely depends on the chemical composition of steel and thermal cycle of welding. For any given grade, the risk of unfavourable structures, such as the occurrence of martensitic structure, can be assessed by the knowledge of the change of the hardness that occurs in relation to the cooling parameter. For any given welding conditions, the microstructure of the high strength structural steels with low carbon and alloy element contents are more favourable.

The toughness of HAZ depends considerably on the welding thermal cycle and on the hardenability of the steel. Toughness is higher when cooling time is short. Nevertheless for any given thermal cycle, we see the favourable influences of the reduction of carbon and nitrogen content on the toughness of HAZs.

The reduction of phosphorus content allows improvement of the HAZ toughness when welding with low heat input especially after a thermal stress relieving treatment. In case of high energy welding, the HAZ toughness of steel with low nitogen content may be improved by adding very small quantity of titanium to improve the microstructure. The reduction of hardening elements has a beneficial effect on the properties of the HAZ and that is why high strength steel result in reduced susceptibility to cold toughness, while leading to a significant reduction of the hardening elements and maintaining high strength.

Improvement of fatigue resistance of weld joints

Fatigue strength of welded joints is improved through a number of techniques in which either weld geometry, weld profile or residual stresses are modified. Effect of improvement techniques on crack initiations lives has been shown in Figure 13.5. Commonly used techniques are as follows:

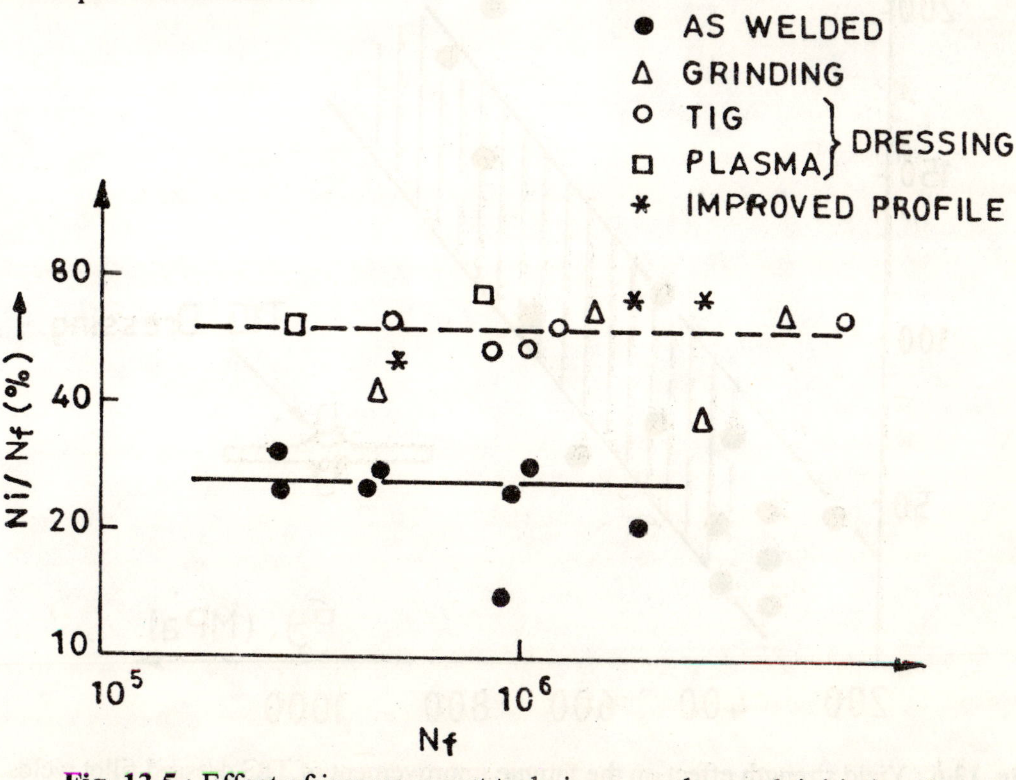

Fig. 13.5 : Effect of improvement techniques on the crack initiation lives

TIG Dressing

TIG welding is used for finishing a weld. The process consists of dressing the toe zone using a TIG torch. Stress concentration is reduced by smoothening the weld profile at the connections and increased hardness in the treated area. TIG finishing can be applied to all types of conventional joints with weld bead over thickness where fatigue rupture initiates in the toe area. The Main reason for the improved fatigue strength of joints welded using TIG finishing is the stress concentration at the weld toe is considerably reduced by smoothening of the connection profile and the defects are decreased by dressing. Yield strength effect on fatigue improvement of TIG dressed fillet welds has been shown in Fig. 13.6. The best results are obtained when finishing pass is performed without using filler metal. Pretreatment scale deposits on the surface to be treated may cause small notches or undercuts along the remelted zone. For this reason, the surface should be cleaned using metal brushes in order to remove scales, slag and other impurities before TIG dressing.

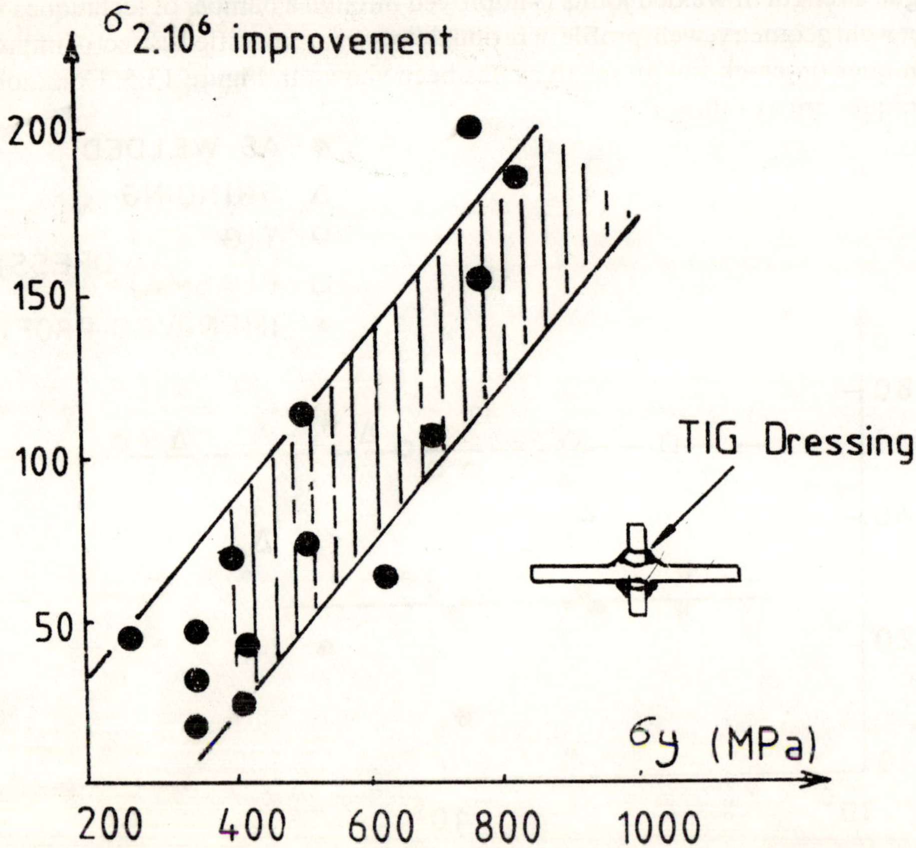

Fig. 13.6 : Yield strength effect on the fatigue improvement of TIG dressed fillet welds.

Plasma Dressing

The principle of the operation is similar to that of TIG dressing. The main difference, in plasma dressing, is that in this twice as much energy is supplied, which enable greater operating speed.

Weld Toe Grinding

This technique is used to eliminate crack initiation by removing metal from the weld zone. This machining can be performed using a disc grinder or rotary burr grinder. To eliminate all slag inclusions and microcracks, machining depth should be 0.5 to 0.8mm. While eliminating crack initiation, this technique enables the weld toe profile to be improved, which reduces the local stress concentration coefficient. The improvement obtained over 2×10^6 cycles varies from 30 to 50%. If the operation is followed by a finishing pass with sand paper, the improvement reaches 80%.

Surface Prestressing

One method for surface prestressing is weld toe hammer peening. This improve technique consists of cold hammering the metal toe using a penumatic hammer. The hammer is equipped with a hemispherical head tool or a needle device. Hammer peening effectiveness depends upon the number of passes and the duration of the operation. If hammer peening is too rapid, the depth of deformed area may be insufficient to surround all defects, with a field of residual compressive strain. Usually an identification depth of 0.6mm is good compromise between treatment time and effectiveness. Shot peening is the other process for surface prestressing in which surface to be treated is sprayed with steel shots. Each impact has the same effect as hitting the surface with a small hammer. The treatment enables a residual compressive stress field of approximately 0.75 yield strength to be introduced in the surface layer.

6. CONCLUSION

- Steels in marine applications are subjected to more adverse conditions as compared to structures at other applications as these are subjected to cyclic loading, temperature variations and corrosive atmosphere.
- Sea water reduces fatigue life by a factor of two compared with results in air.
- Joints made from steels of different strength in general do not differentiate significantly in fatigue strength in the as welded conditions. Fatigue weld strength is almost independent of yield strength.
- Inclusions in steel are to be kept to the minimum and optimum processing parameters should be strictly adhered to.

- The mechanical properties of the weld depends not only on the chemical composition, but also on the microstructure.
- There is marked decrease in fatigue strength with increase of plate thickness.
- Post-Welding processes lead to increased weld strength.
- Weld toes are not the only sites for crack initiation, other causes like lack of penetration, internal defects, too short weld beads etc. are to be eliminated.
- Low carbon steels definitely show improved properties as compared to high carbon steels.
- Higher yield strength steels with low temperature toughness are preferable.
- Developments in steel processing, viz. ladle refining processes, continuous casting, controlled rolling, accelerated cooling, quenching and tempering have lead to marked improvement in quality.

ACKNOWLEDGEMENT

The author is thankful to the management of Research and Development Centre of Iron and Steel, Steel Authority of India Limited for permission to publish this paper.

REFERENCES

1. Steel in Marine Structures, edited by C. Noordhock and J. de Back. Eleevier Science Publishers B.V. Amsterdam, 1987.
2. Product catalogue of Steel Authority of India Limited.
3. Standards from Bureau of Indian Standards.
4. Metallurgy of Iron, volume 12a by P Nilles (Springer - Verlag).

SUBMARINE PIPELINES AND INSTALLATION

14

Lt. Cdr. A. Pande
Deputy Manager (Production) Fleet Maintenance unit, Mumbai

INTRODUCTION

The rapidly depleting land based oil and gas resources coupled with on exponentially increasing consumption to meet the world's sever growing energy demands has forced man to focus on the vast storehouse of resources—the sea. World's major oil reserves have been found in high seas. Northern sea, Gulf of Maxico, Arabian sea etc. have been discovered to hold large reserves of crude oil and natural gas. The production from an offshore oil field is normally transported to shore through oil tankers or through Submarine pipe lines.

In this paper an attempt has been made to highlight the salient features of design, testing, installation, degradation and environmental impact of submarine pipelines. An extensive survey of existing literature has been carried out for this purpose.

Basically Pipeline systems in Petroleum Industry are required for the following purposes:—

(i) gather crude oil from well head,
(ii) transport to refineries,
(iii) distribute the products to process plants, retail markets and to other forms of transportation.

DESIGN OF PIPE LINE SYSTEM

The pipeline and other equipment used to transport crude oil from individual wells and other production units to a central location is called a "gathering system". The design of well fluid transportation and gathering system should takc into consideration the following:—

(a) The well head pressure should be as low as possible to enhance self flow and recovery of fluid
(b) Minimum pressure loss in pipeline
(c) Easy to monitor and control the wells

(d) Minimum loss of hydrocarbons
(e) Accuracy metering well fluid
(f) Flexibility for future expansion
(g) Operational safety
(h) Optimum production cost

Diameter and Wall thickness are the two most important characteristics which influence the cost of pipe line. The diameter determines the friction loss corresponding to the given throughput, hence the energy required. The thickness sets the maximum limit for the operating pressure, and hence the spacing of the pumps along the length of the pipe line.

The design engineer has to estimate the pressure drop and temperature change along the pipeline in order to determine the energy needed to maintain the flow.

FRICTION FACTOR AND FLOW TYPES

The energy balance required for calculation of fluid flow behaviour involves calculation of lost work, the total irreversible energy loss to the inner pipe wall that is unavailable to move the fluid or perform any other action. The term "friction factor" is the empirical term used to assess the numerical term used to assess the numerical value of these irreversible losses, usually called friction losses. The frictional lost work, represented by the frictions resistance, proposed by Fanning is given below:-

$$Wf = \frac{2fLV^2}{gcd}$$

Where;

F = Fanning factor: Experimental have shown that, 'f' is a function of Reynolds number, type of flow and pipe roughness for a Newtonian fluid.

Wf = total friction loss (lost work)

d = diameter of pipe, ft

L = length of pipeline, ft

V = velocity of fluid in the pipeline, ft/sec

gc = gravitational constant,

CALCULATION OF PIPE DIAMETER

The diameter of pipe can be obtained from the following equation

$$d = 1.265q^{0.4}\left[\frac{fLP}{\Delta Pfgc}\right]^{0.2}$$

Where:

q	=	volumetric flow rate
f	=	friction factor
L	=	length
P	=	liquid density
P_f	=	frictional pressure loss
g_c	=	gravitational force

Optimum pipe size is dependent on pipe cost versus pressure drop cost. For non corrosives liquids, the optimum velocity usually is 2.3 m/s (6.10ft/sec) if corrosive liquids like sour amine or glycol are flowing the allowable velocity may be 0.7 - 1.0 ms (2.3ft/s) to minimize erosion corrosion, if carbon steel pipe is used.

CALCULATION OF LINE CAPACITY

Calculation of the capacity of a line with a given pressure drop, is by trial and error method

$$Q = \frac{0.785\, d^2}{P}\left[\frac{\Delta\, Pfgc\ dp}{2fL}\right]$$

Maximum Velocity

Some company specification limit maximum continuous liquid velocity to

$$Vmax = \frac{100A^{0.5}}{P} \qquad ....(8.13)$$

Where;

A = conversion constant and is equal to 1 in English unit.

Flow Improvers

Paraffin deposition takes place as oil cools, while moving up the well and at the surface. The deposits collect in well bore, production tubing and flow line and cause restricted flow which leads increased flow line pressure.

Various methods have been designed to treat high wax crude

(i) Dilution with loss wax content crude
(ii) Pigging of pipelines
(iii) The addition of solutions of oil soluble surfactants i.e. emulsifiers
(iv) The use of wax crystal modifier
(v) Water emulsion
(vi) Reduce energy losses by using drag reducer

STATION DESIGN SPSECIFICATION

Station Design

In designing a pipeline system, the location of pump or compressor stations must be determined as well as the size of individual pumps or compressors within each station. The number and location of stations depends on the length of the pipe line and how much energy must be added to the fluid to transport the required volume at the desired delivery pressure.

Prime Mover Design

The purpose of a pipeline prime mover is to supply the shaft horsepower required by a pump or compressor to move fluid through the pipeline. Type of prime movers includes electric motor, gas turbine and diesel internal combustion engines. Basic selection parameters are horse power output and efficiency. A prime mover must be capable of supplying the shaft horse power required by the pump or compressor and be selected on the basis of initial cost, control adaptability, flexibility of operation and other factors. Calculating the amount of horse power required to drive a compressor involves determining the theoretical horse power required to increase the gas pressure from suction pressure to discharge pressure and the accounting for losses in the compressor. To calculate the break horsepower, the following equation is used.

$$\text{bph} = HQ(\text{sp.gr.})/3960E$$

Where,

Bph = brake horsepower
H = head, ft
sp. gr. = specific gravity of the fluid
E = pump efficiency
Q = liquid flow, gal/min.

INSTALLATION AND TESTING OF PIPELINE

Installation

The installation of a pipeline includes following major steps:

(a) Clearing the right of way as needed
(b) Ditching
(c) Stringing pipe joints along the right of way
(d) Welding the pipe joints together

(e) Applying coating and wrapping to the exterior of the pipe whole length, except a portion of the pipe at each end is sometimes coated before being delivered to the job site).
(f) Lowering the pipeline into the ditch
(g) Backfilling the ditch
(h) Testing the line for leaks
(i) Clearing and drying the pipeline after testing to prepare it for operation

Testing Overview

All completed pipelines must be tested before being put into operation. On long pipelines the line will normally be tested in sections. On short lines the entire pipeline may be tested as a unit. A common approach is hydrostatic testing, ie filling a closed pipeline section with water, then pressurizing the line to a specified pressure to cheek for leaks. Hydrostatic test pressure may be specified as 125% of the maximum design operating pressure of the line.

Non-Destructive testing

Pipeline welds are inspected visually to ensure that they are preformed in accordance with the welding procedure and that the welds are acceptable under the appropriate specifications. Non destructive testing is also required on welds made on a pipeline that will be operated at a pressure that results in a hoop stress of 20% or of the specified minimum yield strength (SMYS). At least 10% the welds made by each welder during each welding day in a liquid pipeline must be non-destructively tested.

LAYING OF PIPE LINES

Requirements For Laying A Pipeline : The submarine pipes are laid on the sea bed. Following are the requirements for laying the pipe line.

(i) Pipe transport vessel
(ii) Lay barge
(iii) Anchor handling tug
(iv) Supply vessel
(v) Diving support vessel
(vi) Survey vessel

The lay barge is most important in pipe laying operation. The following activities are performed on barge:

(a) Pipe handling
(b) Pipe storage

(c) Pipe ends preparation
(d) Line up
(e) Alignment and clamping
(f) Welding
(g) Tension setting
(h) Welding inspection
(i) Field joint coating and filling
(j) Stinger

BURIAL

A pipeline may require burial for the following reasons.

(a) protection.
(b) stabilization.
(c) Thermal insulation.

The aim can be to protect the pipeline from ship anchors, trawl gear, strong currents, high tide range, vortex shedding and/or approach or shallow water. Protections is often imposed by national laws. In other areas protection is generally imposed by environmental conditions. Once the pipeline is laid it can be buried by following techniques:

(a) Burial by jetting method.
(b) Burial by trenching method.
(c) Burial by ploughing.
(d) Simultaneous laying and trenching method.

Where soil conditions prohibit burying the pipeline, mechanical anchoring offers the best safety for the pipe.

CORROSION OF PIPE LINES AND PREVENTION

The submarine pipelines are prone to both internal and external corrosion. The external corrosion results mainly in a loss of metal which causes either a local or general loss of thickness. Internal corrosion may cause the loss of metal but it may also cause an embrittlement of the metal, especially if the effluent flowing through the pipe contains wet Hydrogen sulphide.

To prevent the corrosion and thereby the metal loss, following methods are used:

(a) A special coating together with a cathodic protection system.
(b) Dehydration of the gas.
(c) Chemical inhibition treatments.

Corrosion Mechanism

Sea water corrosion of the steel is of the electrolytic type. Following factors influence rate of corrosion:

(a) Metal composition has negligible effect.
(b) Surface condition is extremely important.
(c) Temperature of sea water plays unfavorable role.
(d) Corrosion rate increases directly with oxygen content.
(e) Suspended solids increase corrosion by an abrasive action on steel surface.

Anti Corrosion Coating

This provides a barrier between steel and sea water so as to form greatest possible electrical insulation. Surface preparation is very important before coating. There are two major groups of coatings applicable to the sealines.

(i) So called "traditional" coatings with hydrocarbon bond (coal tar pitch or petroleum tar): These were used for first sea lines and are still used in many major projects.
(ii) So called "modern" coatings; thermoplastics and thermo hardeners.

There are always composite coatings for sealines, of thickness 5-6mm and made up of: a primer coat, an enamel coat, a reinforcing coat (minimum one) made of glass cloth or fabric, a second enamel coat, and a mechanical protection coat (asbestos felt, glass fabric etc.).

Cathodic Protection

Cathodic protection is an anti corrosion technique which consists in bringing the electro-chemical potential of the metal to a sufficiently negative value so that it is immune to corrosion. The drop in potential required to ensure cathodic protection is obtained by an electric current. The electric current may come from generator or by creating a battery using materials which have different potentials, with the environment forming electrolyte. Protection provided by a corrosion covering is never complete because of the reasons mentioned below and therefore it is essential to supplement the protection system with a cathodic protection system.

(a) Production defects which escaped notice at the inspect.
(b) Damage to the coating during handling and laying of the pipe.
(c) Electrical insulation characteristics of the coating which deteriorates with time (through water absorption aging etc.).

Internal Corrosion

Oxygen, Carbon Dioxide, and hydrogen dissolved in water drastically increase it's corrosivity. Water absorbs oxygen from air, and reacts with iron causing it's corrosion. Similarly carbon dioxide dissolved in water forms carbonic acid and causes pitting of metal known as 'sweet corrosion'. The hydrogen sulphide causes a reaction with steel forming iron sulphide, which sticks to the metal surface called 'sour corrosion'. The free hydrogen also causes hydrogen induced cracking sulphide stress cracking in the pipe line and handling equipments.

POTENTIAL PIPELINE DEGRADATION

Mechanical Damage

The eventual consequence of mechanical damage is pipeline failure and rupture or leak. Causes for mechanical damage may be due to:

(a) Pipeline material, pipeline handling and laying operations or pipeline environment after construction
(b) Pipe steel wall material defect
(c) Weld defect
(d) Badly fabricated anode joint
(e) Dry buckle due to excessive bending and lack of tension during laying
(f) Construction equipment, for instance trenching machines or barge anchors
(g) Flushing equipment such as sea bottom trawlers
(h) Ships anchors impacts or drag, dropped objects
(i) External corrosion
(j) sea waves and current forces
(k) seabed erosion, liquefaction, movements, etc.
(l) Pipeline in service conditions or effluents conditions may degrades due to, internal corrosion, uncontrolled expansions, or bucking due to internal pressure and temperature, transient surges and water hammer.

Mechanical damage may be qualified an external damage if it involves the pipeline shape or external steel pipe surface and coatings. Main types of pipeline mechanical damages are:

(a) Cracks: Cracks (or flow) in pipe wall or inside welds are steel fractures which can be due to various causes listed above. Crack propagation is particularly harmful for high vapor pressure liquids or gas lines for which long running fractures are possible. Equilibrium and propagation of cracks in pipe steel walls can be predicted with the use of fracture mechanics methods provided enough data is available both on the cracks and on the steel material mechanical properties.

(b) Metal loss: Metal loss in pipe wall can be caused for instance by anchor cable friction on an exposed pipeline ripping off external coating and creating a gouge on pipe steel but is generally due to corrosion either internal or external.

(c) Dents: A dent is a geometrical discontinuity on pipe steel wall which may be caused for instance by an impact load.

(d) Buckle: Shell buckles during construction. An ultimate consequence of excessive pipe bending under high hydrostatic is collapse buckling which reduces the internal pipe section to such an extent that two opposite internal pipe wall sides establish contact. After construction pipe shell buckles can only be obtained in shallow depth seas from very high shear forces such as form work barge anchor pipeline dragging, or major impact load (sinking ship).

(e) External Corrosion coating damage: External corrosion coating damage is a risk for small diameter lines which are not coated with concrete for negative buoyancy, and are directly exposed during and after construction, as well as for pipe line risers which are specially protected against external corrosion.

*(f) Degradation of field joints:*Field joint coating condition may be degraded during installation or due to wear. Field joint numbers may be obscured by sediments, or fouling.

(g) Degradation pipeline anodes: Anodes are in the form of Zinc or other sacrificial metal bracelets fitted to the pipeline via welded connections. They are installed at regular intervals along the line, and can be degraded through exhaustion, or become inactive through damage.

Other types of pipeline degradation

Pipe lines degradation, i.e. diminution of strength against operational, environmental or exceptional loads can be due to causes which do not imply a direct damage to the pipeline structure.

(a)Marine growth: Plants algae, kelp and other seaweed can grow from a few centimeter to 1.5 meter long, and may have a positive action in some cases. Marine animals, mussels oysters, tube worm, barnacles, sponges, corals are very different in shape and can be described as either soft or hard marine growth. Growth thickness can be very variable depending on depth temperature and other environmental conditions, and time elapsed after construction, subsequent variations in pipe density and diameter may, if significant, alter pipeline hydrodynamics stability.

*(b) Pipe line free span:*Pipeline free span exceeding design allowable lengths can be a major pipeline degradation which may cause excessive stress, current and waves induced vibrations leading to pipeline fatigue and damage.

(c) Bacterial attacks: Microbial-induced corrosion is a severe problem in many production and water-handling facilities. The two types of bacterial of concern in most cases are sulphate-reducing bacteria and aerobic or slime forming bacteria. Sulphate-reducing or desulfovibrio bacteria are detrimental to systems because they reduce the sulphate ions to sulphide, which reacts with water to form hydrogen sulphide. Hydrogen embrittlement (sulphide stress cracking) or Hydrogen induced cracking may also occur.

(d) Physical variables like pressure, temperature and velocity: Corrosion rates generally increase with temperature. However the highest corrosion rate is generally observed within 40 - 80 degree centigrade range. With pressure more gases go in to solution, which in turn increases corrosivity of the solution Stagnant or low velocity liquids give low corrosion rates. High velocities and or presence of suspended solids can lead to erosion corrosion, impingement or cavitation.

LEAKS

A pipeline leak is not the cause but the consequence of a failure in the chain which links Design, Construction, Operation, Inspection and Maintenance, Repairs and prevention. Therefore any leak should be investigated and corrected in reference with originating failure of the system. In recent years with technological improvements, lot of advanced Leak detection systems have been developed. Leak detection and localization for offshore pipeline is more difficult than on shore because of generally longer distances between eligible checkpoints equipped with instruments and flow meters and because the difficulty to install reliable data transmitting sensor along the sub sea Line.

CONCLUSION

Great advances have been made and continued to be made in submarine pipeline transportation. These include improvement in material and method of pipeline manufacture and better design and construction methods for pipeline and stations. Safe operational procedures, sound construction practices, timely condition monitoring and remedial actions, proper area demarcation and observance of rules and regulation and surveillance will lead to economic prosperity and clean marine environment.

REMOTE MONITORING TECHNIQUE FOR CORROSION PROTECTION MONITORING OF OFFSHORE PLATFORM

15

P. F. Anto

Oil and Natural Gas Corporation Limited Institute of Engineering and Ocean Technology Phase - II, Panvel 410221, Navi Mumbai, India

INTRODUCTION

Corrosion protection of structures using cathodic protection involves the application of current either through galvanic system or through impressing an external voltage. Under these conditions, the structure will be polarized to a potential corresponding to the region of immunity. In this process, the potential of the structure must be maintained and also monitored to avoid damage to the coatings on the structures. Also, the system needs continuous monitoring for ensuring effective protection.

Instrumentation can play an effective role in cathodic protection, especially where impressed current technique is employed for monitoring and controlling the protection potential. Even though the impressed current technique, is straightforward and simple, the huge size and complexity in shape of the structure and the need for desired protection level at all places make the instrumentation part more complex demanding careful sacrificial anode technique which does not require instrumentation once the efficiency of the anode is established.

Various non-automatic and automatic devices are used in cathodic protection for providing adequate protection to the metals in service. In the case of control system various technological developments have been reported. They are :

a. Rectifier and auto transformer control in non-automatic modes, and
b. Motor operated variac, magnetic amplifier control, transistorized potentiostat, silicon controlled rectifier, microprocessor based system control etc., in automatic modes.

Various recent developments have taken place in view of the availability of modern electronic gadgets at a lower cost and the flexibility of introducing such system with visual recording and long term data logging. In this connection a Supervisory Control And Data Acquisition (SCADA) with remote monitoring possibility through P&T networks, has been developed.

Any huge structure like an offshore platform or a dry dock has a number of non-symmetrical structures to be protected. Because of engineering limitation, non-uniformity gets introduced either in the location of anodes or in the location of reference electrodes. The requirement of an intelligent protection system in a multianode and multireference electrodes situation is the problem of the day.

SYSTEM DEVELOPED FOR OFFSHORE STRUCTURE

The present control system is designed using microprocessor based closed loop control system. The system is designed in such a way that it includes the following points:

(1) The output current of the rectifier is impressed in the form of periodic pulse with constant amplitude.

(2) The duty cycle of the pulse is proportional to the potential difference between the corrosion free set potential and true structure potential.

(3) The structure potential is measured during the off period of the current pulse which measures the true structure potential free from the IR drop of the medium in between the structure and reference electrode.

(4) The pulse rate is fixed depending upon the relaxation time of the medium.

A personal computer is used to monitor the status of the cathodic protection system from the control room. The communication unit is designed in the serial mode operation which is based on asynchronous digital communication technique. A telecommunication network is used to transfer data and commands between the monitoring PC at the control room and the remote control unit which is located the nearer to the structure under CP.

SYSTEM HARDWARE

The system hardware comprises of the following:

1. Central Processing Unit (CPU)
2. Input/output ports of Programmable Peripheral Interface (PPI).
3. Communication Hardware.

CENTRAL PROCESSING UNIT

A CPU includes the microprocessor and additional components like memory devices, buffers, decoders, and clock-drivers. The main work of the CPU is to execute the program stored in the memory and generate suitable signals to their input and output ports.

INPUT DEVICE

In order to provide more effective cathodic protection, one has to ensure true reference electrode potential measurement free from any noise signal resulting from stray current distribution, industrial noise and high power transmission system disturbance etc. With effective filters (Low Pass) and DC level offset adjustment input buffer amplifier along with a multi-channel multiplexer the CPU is able to measure appropriate reference electrode potentials.

OUTPUT DEVICE

The output control unit of the SCADA system incorporates an output port of a programmable peripheral interface (PPI), a solid state relay (SSR) to ensure complete isolating among the sophisticated hardware, the rectifier power source and the rectifier unit.

COMMUNICATION DEVICE

The control system based on microprocessor is a parallel device. It transfers eight bits of data simultaneously over eight data lines. For a long distance, parallel data communication is more expensive and hence asynchronous serial communication mode is used, whereby one bit at time is transferred over a single line.

SYSTEM SOFTWARE

The system software of the automatic control system has been developed for monitoring the status of the structure under cathodic protection from the control room which is located remotely from the structure.

SCADA software has three major functions. First, it allows user to monitor the structure potential at various locations. Second, it helps to display the results of the structure potential in the form of equi-potential mapping. Finally, it helps to diagnose the problem of the CP system.

As far as the range of distance between the control room and the structure is concerned, with its powerful communication software package, it can be used for any remotely located structure.

It offers useful record of periodic data corresponding to the status of the structure under cathodic protection. With this data logging feature, it provides appropriate alarm indications to alert the user with regard to any system failures. It can also analyse the corroding nature of the structure.

FIELD TRIAL

This unit was installed in one of the process platforms of Bombay offshore where the platform jacket is protected by sacrificial anode system. With the help of this unit, protection potential on the jacket and current output of anodes were monitored. The system had been found to work satisfactorily. Details are shown in Figs. 15.1 and 15.2.

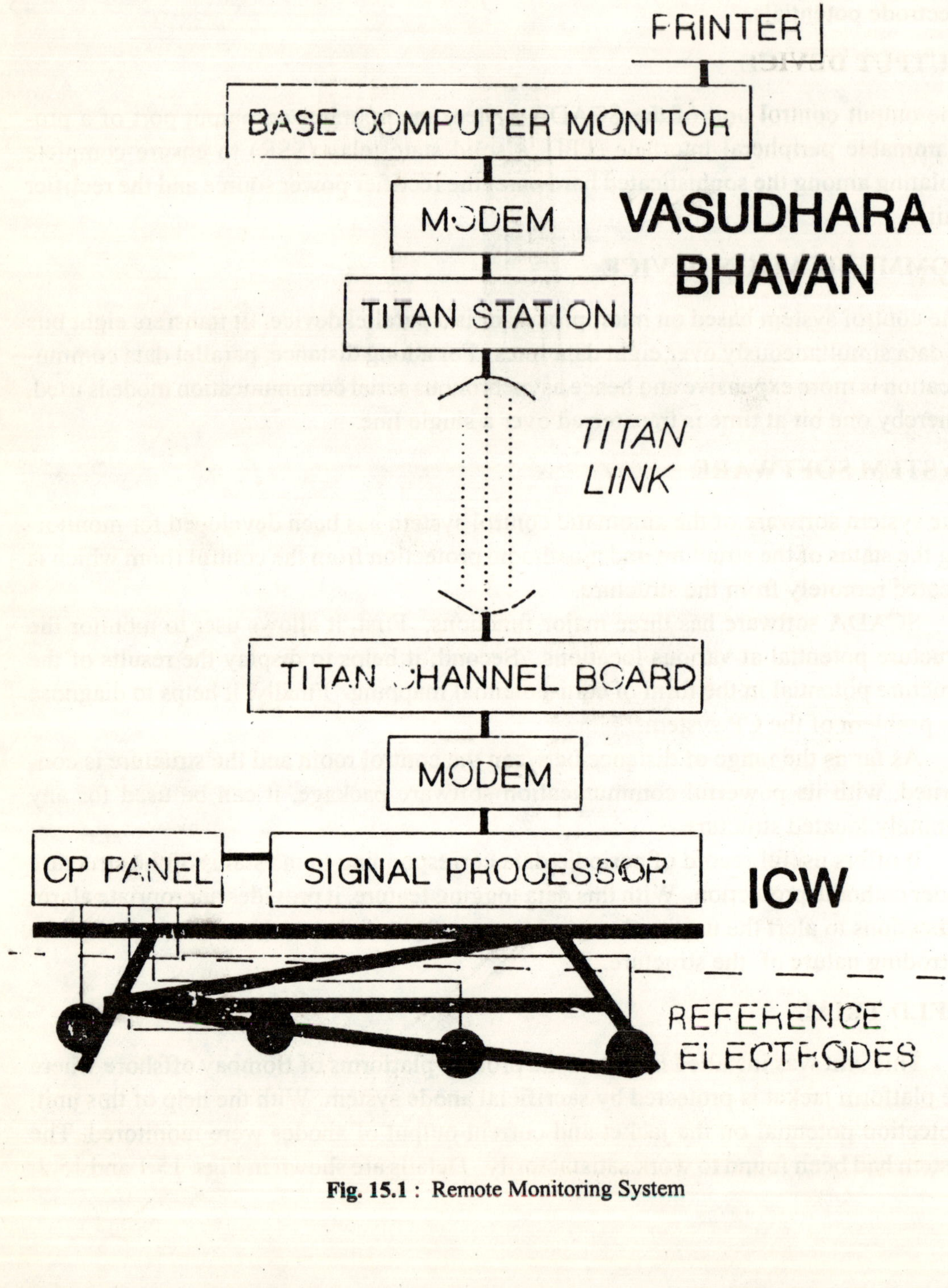

Fig. 15.1 : Remote Monitoring System

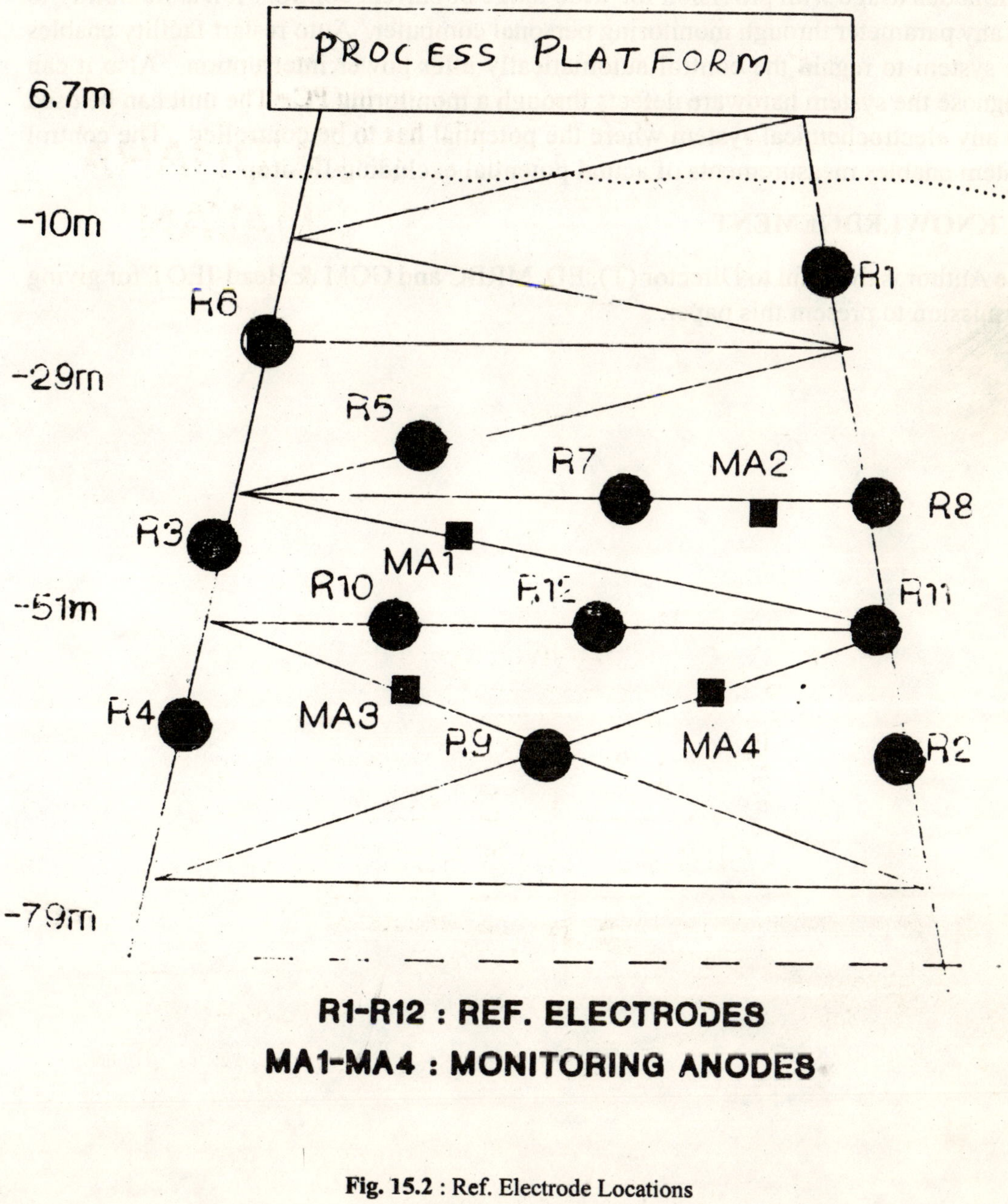

Fig. 15.2 : Ref. Electrode Locations

CONCLUSION

The SCADA control system uses an existing telecom line for data communication between the RTU and monitoring PC. The unit has data logging facility. All the components for this control system can be easily fabricated. The unit is reliable for continuous usage with provision for wide range of current control. It has flexibility to set any parameter through monitoring personal computer. Auto restart facility enables the system to regain the control automatically after power interruption. Also it can diagnose the system hardware defects through a monitoring PC. The unit can be used for any electrochemical system where the potential has to be controlled. The control system enables measurements of actual potential excluding IR drop.

ACKNOWLEDGEMENT

The Author is thankful to Director (T), ED, MRBC and GGM & Head-IEOT for giving permission to present this paper.